Bronzeworking
on Late Minoan Crete

A diachronic study

Lena Hakulin

BAR International Series 1245
2004

Published in 2016 by
BAR Publishing, Oxford

BAR International Series 1245

Bronzeworking on Late Minoan Crete

ISBN 978 1 84171 605 3

BAR Publishing is the trading name of British Archaeological Reports (Oxford) Ltd.
British Archaeological Reports was first incorporated in 1974 to publish the BAR
Series, International and British. In 1992 Hadrian Books Ltd became part of the BAR
group. This volume was originally published by Archaeopress in conjunction with
British Archaeological Reports (Oxford) Ltd / Hadrian Books Ltd, the Series principal
publisher, in 2004. This present volume is published by BAR Publishing, 2016.

Printed in England

BAR
PUBLISHING

BAR titles are available from:

        BAR Publishing
        122 Banbury Rd, Oxford, OX2 7BP, UK
EMAIL   info@barpublishing.com
PHONE +44 (0)1865 310431
  FAX   +44 (0)1865 316916
        www.barpublishing.com

Table of contents

List of figures

Acknowledgements

The present monograph is a revised version of my MA Thesis at the University of Helsinki, Finland.
I would like to thank Professors Ari Siiriäinen and Mika Lavento for all their advice and encouragement and particularly my supervisor, Professor Carole Gillis, University of Lund, Sweden for her help and guidance. The interpretations are, however, my own, as are any errors. Thanks are also extended to Raimo Haapakorpi for his computer assistance and to Margot Stout Whiting for English-language editing. I would also like to thank my companion in exploring Crete for twenty years, Jill Aschan, for continuous, stimulating discussions on the Minoan culture. A special thank goes to my husband for helping me in metallurgical matters and for his long patience with my studying. This work is dedicated to him.

Chapter 1. Introduction

1.1 Background

Access to metals and metallurgical skills were crucial factors in the emergence of the palatial cultures during the Bronze Age in the eastern Mediterranean. Knowledge of Bronze Age metallurgy and its development is thus one prerequisite for the understanding of these cultures. The high standard of the bronzeworking on Late Minoan Crete is well known and attested by technically sophisticated and artistically beautiful weapons and vessels found in the Minoan and Mycenaean palaces and tombs. In the monumental building programmes on Crete, the availability of a wide range of effective bronze tools was essential.

Many important questions arise. Where did the Minoans get their metal? There are virtually no copper ores on Crete and tin is barely available in the whole Aegean. All the metal needed for the large-scale bronze industry had thus to be imported. Where did their metallurgical skills and technology emerge? Today we assume that the technology originally came from the Middle East or Anatolia. But on the other hand, according to the Greek myths, copper metallurgy originated on Crete, Hephaistos is mentioned in the genealogy of the Cretan Kings and Phaistos was the centre of copper metallurgy. Myths should perhaps not be totally neglected. Scientists are indeed faced with many questions which hardly ever can be satisfactorily answered.

Surprisingly few studies on Minoan bronzeworking have been published, if compared with studies on prehistoric metallurgy on the Aegean islands and Cyprus. One reason may be the lack of ore sources on Crete. Thirty years ago Keith Branigan published two pioneering holistic studies *Copper and Bronze Working in Early Bronze Age Crete* (1968) and *Aegean Metalwork of the Early and Middle Bronze Age* (1974). Since then, no synthesis of Minoan metalworking has been published; research has been focussed on typological studies of some types of bronze objects which traditionally have been the hallmark of Minoan bronzeworking. But in a recent study *Minoan Crafts: Tools and Techniques. An Introduction. Vol I and II* (1993, 2000), R.D.G. Evely has thoroughly investigated the bronze tools and the metalworking technique.

Evidence for metallurgical activities was rarely noticed in older excavations. However, during the last twenty years crucibles, moulds, slag and other evidence for bronzeworking have been found in many Late Minoan excavations, *e.g.,* in the Unexplored Mansion at Knossos, and in Kommos,

Mochlos, Poros-Katsambas and Palaikastro. Important evidence for a Protopalatial bronze workshop has also been found in the industrial area Quartier Mu at Malia. The metal and metallurgical evidence from these projects have been analyzed by specialists and published in separate reports. But the whole process, including raw materials, bronzeworking and finished objects have been little studied by scientific methods. A serious problem has been that Greek museums have not been willing to lend out objects for scientific investigations. Chemical analyses published even in recent publications are thus often old and no longer reliable.

Perhaps the most interesting question concerning bronzeworking on Late Bronze Age Crete is the origin of the Minoan copper. Previously the generally accepted, but unproved, hypothesis was that it came from Cyprus. But in the early 1980s, Noel Gale and Zofia Stos-Gale could, with lead isotope analyses, convincingly prove that the main source of the copper in Late Minoan bronze objects was Laurion (Gale and Stos-Gale 1982). The majority of the copper oxhide ingots, however, remain an enigma; their source has so far not been identified. The whole problem complex related to the Minoan copper sources is still unsolved, but actually no one has seriously tried to do it.

In summary, so far no holistic study and synthesis of Late Minoan bronzeworking has been published. This study is the first attempt to collect all available data and analyze the whole bronzeworking process from raw materials to finished objects and present a comprehensive view of its development from the Neopalatial to the Postpalatial period. The approach in the study is techno-economical.

1.2 Aims of the study

The aims of the study are

- to collect all published data on bronze objects, evidence for raw materials and bronze workshops as well as scientific analyses related to Late Minoan bronzeworking, and store them in codified databases,
- to analyze this information and identify differences in the bronze object assemblages, selected object types, the finding contexts, the metalworking techniques, the alloy composition and the availability of copper between the Neopalatial, the Mycenaean Knossos and the Postpalatial periods,

- to identify the causes of possible changes and differences between these periods,
- to present an overview of the bronze industry on Crete and its development during the Late Bronze Age.

If any diachronic development in the Minoan bronze industry from the Neopalatial to the Postpalatial period can be identified, the main question is: what does it indicate? Changed priorities or new uses of the bronze? Improved or debased metalworking technique? Changes in the availability of copper and tin? And how are the dramatic social events on Crete reflected in metalworking?

1.3 Scope of the study

Metallurgically and geographically, the scope of the study is limited to copperbased objects, archaeological evidence for their manufacturing and copper raw material found on Crete, as well as scientific analyses related to them. Lead and precious metals are not included as their social role was minor compared to bronze and their manufacturing techniques different. I have not discussed the complex, unsolved problem of the tin sources nor the Minoan trade with bronze objects. Minoan objects found outside Crete are thus not stored in the databases. Chronologically the study is limited to the Late Bronze Age.

The many codified parameters enable a wide range of analyses by freely searching and combining parameters and values in the databases. For this study, I have, however, chosen to only compare the data for the following three periods defined by Evely (1993: xxiii) and to try to identify possible changes and developments in bronzeworking during the Late Minoan period.

- The Neopalatial period (MM III - LM I or 1700 / 1600 - 1450 BC)
- The Mycenaean Knossos period (LM II - LM IIIA1 or 1450 - 1375 BC)
- The Postpalatial period (LM IIIA2 - LM IIIC or 1375 - 1050 BC)

1.4 Contents of the study

As an introduction to the topic, Chapter 2 contains a brief outline of the Early and Middle Minoan bronzeworking. The information sources used for collecting the data and the structures of the databases are presented in Chapter 3. The Appendices II - V comprise the database parameters, abbreviations and printouts of the entire databases. The data analyses presented in Chapter 4 are based on bar charts and curves produced from the databases. As a result, a preliminary overview of the development of the Late Minoan bronzeworking is presented in Chapter 5 and the summary and conclusions in Chapter 6.

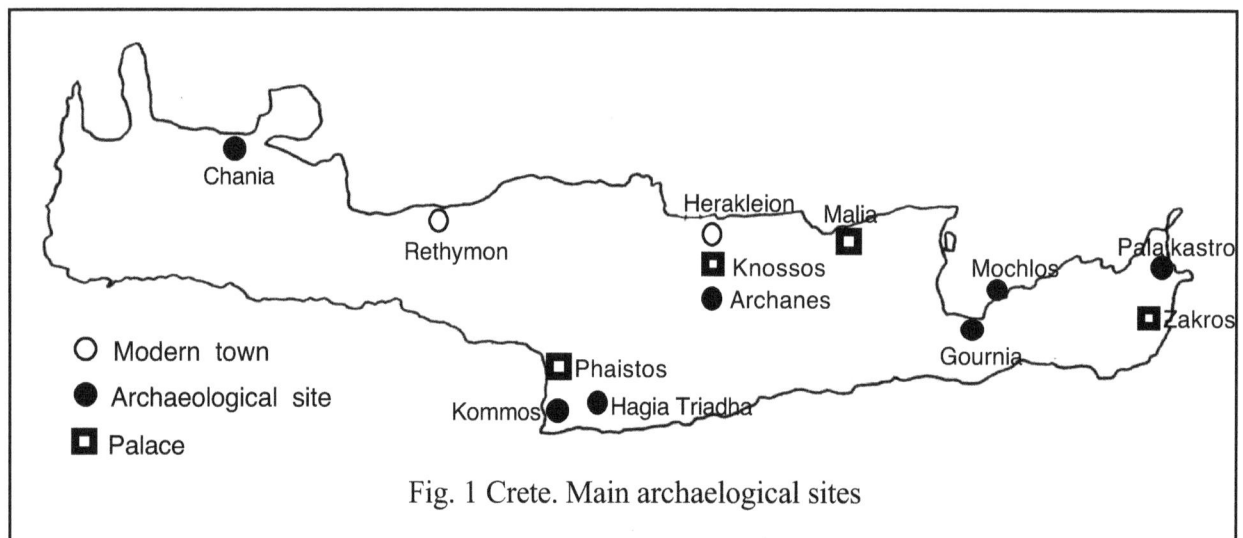

Fig. 1 Crete. Main archaelogical sites

2

Chapter 2. Early and Middle Minoan bronzeworking

2.1 Early Bronze Age

The focus of prehistoric metallurgical studies on Crete has always been the Early Bronze Age, probably partly initiated by Colin Renfrew's famous *Emergence of Civilisation. The Cyclades and the Aegean in the Third Millenium B.C.* (1972). The main questions have been: Where did Minoan metallurgy originate? Why and how did Crete develop into a metallurgical centre in EM II? How did the contacts with the Aegean develop? Which were the copper sources used, and are the prehistorically accessible ore sources exhausted? And especially - were Crete's own copper sources exploited and sufficient or had the copper to be imported?

Keith Branigan's two studies (1968, 1974) were pioneering and are still the only comprehensive reference works on Minoan bronzeworking. His catalogue of Cretan Early Minoan bronze objects contains about 500 objects of which almost half are daggers and the rest a wide range of tools and ornaments (Branigan 1968: 71-99). The bulk of these finds were, however, recovered from the Mesara tombs a hundred years ago, and as the tombs were used for generations, the dating has been problematic. During the last fifty years substantial finds of Early Minoan metalwork have been frustratingly few (Branigan 1974: 4). The important finds from the cemetery of Agia Photia are still not published. The alloys used were mainly arsenical bronzes, either deliberately produced alloys or accidental alloys from arsenious cupriferous ores. The share of tin bronzes was low (Branigan 1974: 71-74).

The possible use of the Cretan copper ores has been a vexing question. The evidence is scanty. Branigan first believed that Cretan ores were used and were sufficient, but later changed his mind. In the 1980s, convincing evidence for copper mining and smelting in the Early Bronze Age was found on the island of Kythnos. The ore has been proved to have been the main copper source for the Early Minoan bronze objects (Stos-Gale *et al.* 1988; Hadjianastassiou and MacGillivray 1988; Stos-Gale 1989, 1998). Recently, the first Bronze Age copper smelting furnace on Crete, dated to EM III, has been securely identified at Chrysokamino in the Kavousi area in eastern Crete (Betancourt *et al.* 1999; Betancourt 2001; Betancourt *et al.* 2003). Excavations

at the site have yielded various metallurgical remains such as slag fragments, amounting to approximately 700 kgs, abundant fragments of coarse ceramics from a cylindrical shaft furnace bearing perforations and fragments of pot-bellows. The major technical parameters of the smelting operation were investigated by an analytical programme for the metallurgical remains. A smelting furnace of a similar type as that of Crysokamino has recently been discovered on Kythnos which indicates a particular copper smelting technology in the southern Aegean region (Bassiakos and Betancourt 2002). Due to these important finds and several promising new attestations for prehistoric mining on the Cycladic islands, the Early Bronze Age metallurgy in the southern Aegean is presently being re-evaluated.

Recently, a new programme on Minoan metallurgical techniques has begun with the aim to investigate copperbased artefacts in combination with their manufacturing technique. The prerequisite for the project is that the museums in Herakleion and Chania are willing to lend their objects for analytical and metallographic investigations. So far only the results for 10 objects have been published, almost all from the Early Bronze Age. The results show that two types of manufacturing were employed: either casting and coldworking or casting, annealing and coldworking. (Tselios 2003)

2.2 Middle Bronze Age

According to Branigan, the high standard of metalworking in the Prepalatial period continued during the Protopalatial period, and Crete had a central role in the development of the Middle Bronze Age metal industry in the Aegean. The major weapons of the Late Bronze Age evolved and several types of standard Late Minoan equipment were adopted, *e.g.,* the double-axe (Branigan 1974: 114-119). Surprisingly few bronze objects from securely dated Middle Minoan contexts have, however, been found. A considerable number of them come from Malia. In Quartier Mu, the industrial area of Protopalatial Malia, important evidence for a bronze workshop has been found, which strongly supports the hypothesis of Malia as a metallurgical centre. The evidence shows that arsenical bronzes were already replaced by tin bronzes. (Poursat 1996: 45-57, 115-118)

Chapter 3. Databases

All data in the study are published information, drawn from a wide range of sources. They have not been checked or changed even if they might no longer be reliable. This is true for object classification and particularly for the chemical analyses and the dating. They are stored "as published" in the databases. These published data are as such sufficient for a first, general overview of the bronzeworking. It is to be hoped that more accurate data will be available in the future. My intention is to continously update the databases. The data are stored in four databases codified with appropriate parameters. (App. II-V) in a Macintosh environment under the operating system Mac OS 9.1. The database and spreadsheet programmes used are Appleworks K 1 - 5.04.

3.1 Information sources

The data on bronze objects is primarily collected from catalogues, of which the most important are Evely's study on Minoan tools and techniques (1993; 2000), Matthäus' study on Aegean bronze vessels (1980), Kilian-Dirlmeier's study on Aegean swords (1993) and Sapouna-Sakellarakis' study on Minoan figurines (1995). The information in the catalogues is supplemented by newer finds and analyses. Several object types are, however, not compiled in catalogues, e.g., daggers, knives, razors and ornaments. The data for these object types are collected from excavation reports for, e.g., the "Warrior Graves" in the Knossos area (Evans 1906, 1914: Forsdyke 1926-27; Hood and de Jong 1952; Hood 1956; Catling and Catling 1974) and Gournia (Boyd Hawes 1908). Other important sources used are Boardman's study on the objects from the Dictaean Cave (1961), Catling's study on Cypriote bronzework, which also contains data on Minoan objects (1964), Georgiou's study of "metal groups" in LM I destruction levels (1979) and E. M. Platon's unpublished dissertation on the bronze tools from Zakros (1988). Two studies covering the whole of Crete contained useful information on bronze objects: the Gazetteer of Neopalatial sites in *The Troubled Island* by Driessen and Macdonald (1997) and Kanta's study on LM III Crete (1980).

A serious problem in the data collection has been that the documentation for several important excavations is insufficient or even missing. Hagia Triadha, for instance, with important bronze finds, has from the very beginning been inadequately documented (Halbherr *et al.* 1980). Some recent important excavations have not yet achieved a final publication, *e.g.,* Archanes and the LM III cemetery Armenoi. The information from the most recent excavations is collected from annual excavation reports, *e.g.,* for Mochlos (Soles and Davaras 1992, 1994, 1996).

Few new chemical analyses of bronze objects have been published. The main sources have thus been the catalogues where older analyses are collected. Sapouna-Sakellakis' study is an exeption, with 45 new analyses of bronze figurines (1995). The most recent publications are Mangou's and Ioannou's chemical analyses of prehistoric Greek bronze objects and copper ingots (1998, 2000). Tselios has so far analyzed only Early Minoan objects (2003). Other sources used are Catling's and Jones' tin analyses of objects from the Unexplored Mansion (1977) and Northover's analyses of objects from the Ashmolean Museum (1995).

The data for copper ingots were found in different publications. Evely's study (2000) and Mangou's and Ioannou's article (2000) contain dimensions and chemical analyses but the interesting proveniences for the copper in the ingots and also in artefacts are published by Gale and Stos-Gale in a great number of articles. The results are somewhat varying and no compilation and state of the art for the material from Crete has been published. For this reasons the provenience of the copper is only discussed on a general level.

Evidence for metallurgical activities was normally not noted or documented in early excavations, but during the last twenty years, such evidence, *e.g.,* crucibles, moulds, slag and casting debris have been found on many Late Minoan sites. They are published by specialists in separate reports. The most important are from the Unexplored Mansion at Knossos (Catling and Catling 1984), Kommos (Blitzer 1995), Palaikastro (Hemingway 1996) and Chania (Stos-Gale *et al.* 2000). Final reports from the metallurgically important sites of Mochlos and Poros-Katsambas are so far not published. The recent Vol. II of Evely's study *Minoan Crafts: Tools and Techniques. An Introduction* (2000) contains rather complete catalogues of crucibles, moulds and other indications of metallurgical activity.

3.2 Bronze objects

The database for bronze objects contains the following 24 codified parameters for each object:

- findspot, 5 parameters (site, context, geographical area, location and site type)
- dating, 2 parameters (period and date)

- object type, 1 parameter
- category of object, 1 parameter
- typologies for some object types, 3 parameters
- condition, 1 parameter
- measurements, 5 parameters (length, width, height, rim diameter, weight)
- contents of main metals, 4 parameters (Cu, Sn, As, Pb)
- museum or museum number, 1 parameter
- references, 1 parameter

The parameters and their codifyng are my own. They are presented in Appendix II.1, the abbreviations for the sites and their codifying in Appendix II.2 and the abbreviations for the museums in Appendix II.3.

All information in the database is published data. If varying values occur in the literature, the most recent data are used. Uncertain datings as, *e.g.*, LM I or LM III, I have classified as Unspecified LM. If the data are collected from a catalogue, I refer only to the catalogue due to space limitations. More detailed references to, *e.g.,* excavation reports can be found in the catalogues. Additional references are stored in the database only if they contain information supplementary to the catalogue.

The database contains data for 1857 bronze objects found in Late Minoan contexts on Crete. As the manufacturing site for the objects in general is unknown, no separation is made between objects manufactured on Crete and imported objects. The database is restricted to objects and evidence found on Crete, and does thus not contain Minoan bronze objects found outside Crete. Many objects do not have a complete set of parameters: supplementary information, if available, can be added later.

The aim of the study has been to build up databases which are as complete as possible. The main problems in the data collection have been incomplete or missing excavation reports and the, considerable number of bronze objects in private collections (*e.g.* Kontorli-Papadopoulou, 1984). An example is the Mitsotakis collection which was exhibited in the Cycladic Museum in Athens. The catalogue contains about thirty Late Minoan bronze objects (Μαραγκού 1992: 233-267). Presently the objects are in the Chania Museum. These objects are not stored in the database, nor have all the most recent Greek journals been scanned. A rough estimate is, however, that the database comprises 75-80 % of the published Late Minoan bronze objects.

A printout of the database for bronze objects sorted by object type, period and site is presented in Appendix V.2 .

3.3 Chemical analyses

Published chemical analyses are often not clearly connected to a particular object and cannot thus be included in the bronze object database. Hence all chemical analyses for bronze objects and metallic evidence from workshops are collected in a separate database which contains 302 analyses. The information compiled is:

- site, 1 parameter
- dating, 2 parameters (period and date)
- object type, 1 parameter
- content of main metals, 4 parameters (Cu, Sn, As, Pb)
- method of analysis, 1 parameter
- date of analysis, 1 parameter
- museum or museum number, 1 parameter
- references, 1 parameter
- number in object database, 1 parameter

A printout of the database sorted by object type, period and site is presented in Appendix V.1.

3.4 Copper ingots

The database for copper ingots contains the following 27 codified parameters:

- findspot, 5 parameters (site, context, geographical area, location and site type)
- dating, 2 parameters (period and date)
- type of ingot, 1 parameter
- condition, 1 parameter
- measurements, 4 parameters (length, width, thickness and weight)
- content of main metals, 10 parameters (Cu, Sn, Pb, As, Sb, Fe, Ni, Co, Zn, Bi)
- total metal content, 1 parameter
- provenience, 1 parameter
- museum or museum number, 1 parameter
- references, 1 parameter.

The coding of the parameters is the same as for the objects (App. II.1, 2, 3). The database comprises 75 ingots and ingot fragments. A printout of the database sorted by ingot type, condition and period is presented in Appendix III.

3.5 Metallurgical evidence

Of the common material remains for on-site working of metals (Fig. 2), only the data for crucibles, moulds, tuyéres and pot bellows are collected in databases. The evidence is not large and are hence collected in manual tables. The few finds from the Proto- and Prepalatial periods are also included in the appendices to give a better view of the whole development during the Bronze Age .

The information is collected in four tables (App. IV.1 - IV.4), of which the tables for crucibles and moulds are further divided by type. The information compiled is

- site
- dating
- material
- dimensions
- condition / description
- museum number
- references

In all, the tables comprise information of 66 crucibles, 69 moulds, 18 tuyères and 4 pot bellows from a total of 12 workshops. The recent important finds from Poros-Katsambas are not included nor are the large number of hemispherical bridge-spouted crucibles from a Neopalatial context at Palaikastro (Evely 2000: 349) as no detailed information has been available.

Other attestations of metalworking such as furnace remains and fragments, slag, burnt spots as well as metal scrap and waste are not treated quantitatively except for the 52 analyses of the tin content in metallic metallurgical evidence from the Unexplored Mansion. They are included in the database for Chemical analyses (App. V.1). Unfortunately, slag from Late Minoan Crete has never been investigated and published.

3.6 Bronze workshops

Brief outlines of the eight most important bronze workshops (Malia, Gournia, Mochlos, Zakros, Poros-Katsambas, the Unexplored Mansion at Knossos, Kommos and Palaikastro) and their metallurgical evidence are presented in Appendices I.1 - I.8.

Evidence / Site	Ingots	Scrap	Crucib.	Moulds	Tuyéres	Bellows	Metal bars, strips etc	Waste run-off	Slag	Tools	Furnaces, hearths
Chrysokam.	-	-	-	-	-	•	-	-	•	-	•
Malia,Qu.Mu	x	-	x	•	•	-	-	-	-	-	-
Malia	-	-	-	•	-	-	-	-	-	-	-
Gournia	•	x	x	•	-	-	x	-	-	-	-
Mochlos	•	•	-	x	-	-	x	x	x	x	-
Zakros	•	-	x	x	-	-	-	-	-	-	-
Poros/Kats.	•	x	•	x	x	x	x	x	x	x	•
Knossos, UM	-	•	•	x	-	-	•	•	x	•	x
Kommos	x	x	•	•	-	•	•	x	•	x	•
Palaikastro	-	-	•	•	•	-	-	-	-	-	-
Chania	x	x	x	x	-	-	x	x	x	-	-

Note. The most important evidence on the site are marked by •

Fig. 2 Material remains as evidence for on-site working of metals on Bronze Age Crete.

Chapter 4. Data analyses

The aims of the analyses of the collected data are to identify possible changes in metalworking from the Neopalatial to the Postpalatial periods. This is done through comparing diagrams produced from the databases separately for the Neopalatial, Mycenaean Knossos and Postpalatial periods concerning bronze object assemblages as well as dimensions and alloys for selected object types and the types of metalworking equipment. Based on these results, I draw some general conclusions concerning the bronze industry during the different periods and present some possible causes for the changes.

4.1 Bronze objects

4.1.1 Introduction

The 1857 objects stored in the database originate from 137 different sites (App. II.2), but the major part of them were concentrated on some few important sites. The top ten of these, by number of finds, are presented in Figure 3. Of the total number of objects, 58% were found on these ten sites. Both the type of the sites and their finds vary considerably. In the Psychro Cave, with 155 finds, the majority are cult objects and weapons whereas tools dominate in, e.g., Zakros (151 finds) and Gournia (145 finds). In Mochlos (72 finds) and Sellopoulo (55 finds) vessels constitute a considerable part of the assemblages.

According to published information, only 700 of the objects in the database are situated in the Herakleion Archaeological Museum, but in practice the amount may be almost double, as most of the 854 objects without museum number are probably in Herakleion. According to present directives, the finds have to be placed in local museums, e.g., in Chania, Rethymno, Hagios Nikolaos and Siteia. Only 10 % of the objects are in foreign museums, the majority in the Ashmolean Museum in Oxford (127 objects). But of my catalogue sources, only Sapouna-Sakellarakis (1995) seems to have investigated private museums and collections outside Greece. The other catalogues are mainly concentrated on finds from the public museums. A well-known problem is that large collections of unpublished finds without secure dating and provenience are in private hands and thus normally missing from the scientific literature.

4.1.2 Object assemblages

The entire Late Bronze Age
The structure of the bronze object assemblage for the entire Late Bronze Age is presented in Figures 4-7. Almost half of the objects (853 objects or 46%) are dated to the Neopalatial period. Additionally, most of the objects with uncertain dating (299 objects) probably belong to this period. The finds from the Mycenaean Knossos (443 objects) and the Postpalatial periods (262 objects) are considerably less numerous (Fig. 4).

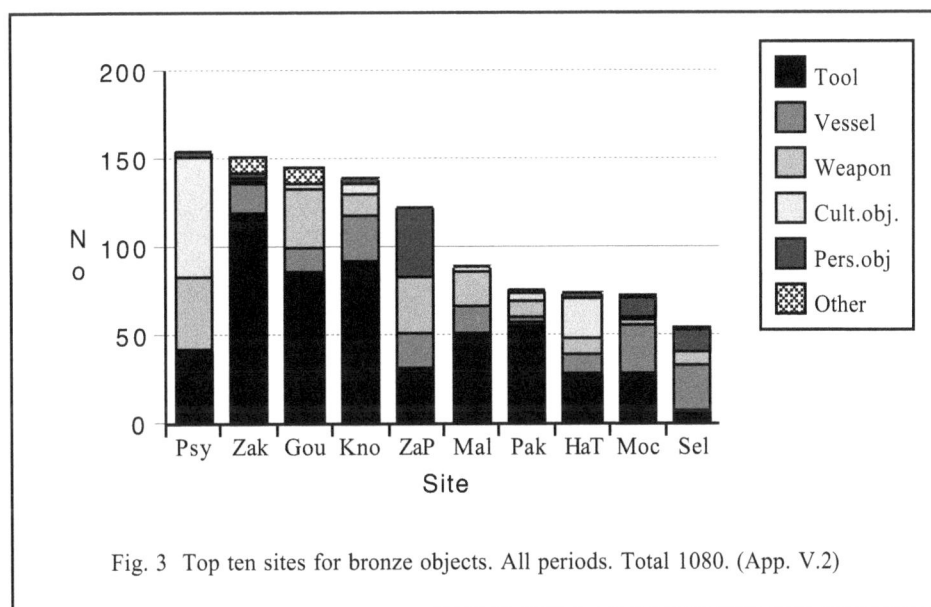

Fig. 3 Top ten sites for bronze objects. All periods. Total 1080. (App. V.2)

Tools constitute the bulk of the objects (878 tools or 47% of the total), including both sizable, heavy duty tools such as double-axes, adzes, chisels and saws for tree felling as well as smaller tools for precision work. (Fig.5) In total, 220 bronze vessels, considered typical of both the Late Minoan and the Mycenaean cultures, are found. The weapon group includes daggers, swords, spear- and arrowheads and the only bronze helmet found on Crete. All knives are classified as tools, even if the long knives found in tombs during the Mycenaean Knossos and Postpalatial periods in reality were weapons. In total, 366 weapons, or 20% of all objects, were found. The 188 cult objects (10% of all) are votive double-axes and figurines. The number of personal objects, mainly razors, mirrors and ornaments, are of the same order of magnitude; 170 in total.

The majority of the finds originate from tombs, in total 584 objects or 31% (Fig. 6). Almost all of them are dated to the Mycenaean Knossos or the Postpalatial periods, as very few Neopalatial tombs have been found. Most of these are from the Knossos area. The total number of settlement finds are, however, higher (856 or 46%), of which 394 come from palaces and 462 from towns and other settlements. The majority are dated to the Neopalatial period and originate to a great extent from bronze collections. Two thirds of the palace

finds come from villas and towns in the surroundings of the palaces. Except for Zakros, the palaces were all destroyed and probably already robbed in antiquity. The most numerous collections of bronze objects have, however, been found in cult caves, particularly in Psychro and Arkalohori. Of the many finds from the peak sanctuary Iuktas, all are not included in the database due to lack of adequate information.

The regional distribution of the finds shows that equal amounts of bronze objects have been found in eastern Crete (740 objects)) and around Knossos (727 objects) or together more than three quarters of all objects (Fig. 7). The finds from eastern Crete originate mainly from settlements and are dated to the Neopalatial period while a large share of the finds from the Knossos area come from tombs. The "Great Minoan Triangle" on the Mesara plain has yielded 184 finds, the main part coming from Hagia Triadha. Kommos is interesting mainly due to the abundant finds of metallurgical evidence (App. I.7). The finds from western Crete are, as expected, few but the finds from the large LM III cemetery in Armenoi are not yet published.

The Neopalatial period (MM III - LM I)
The distribution of the 853 objects from the Neopalatial period concerning object type, site type and geographical region is presented in Figures 8-10.

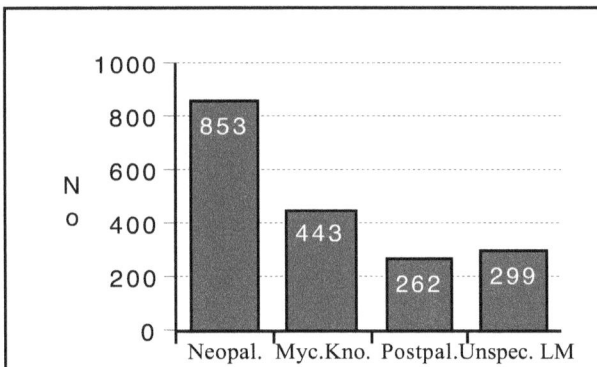

Fig.4 Bronze objects by period. Total 1857. (App. V.2)

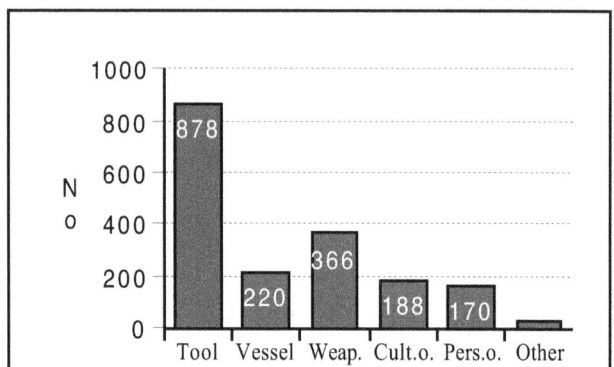

Fig. 5 Bronze objects by object type. Total 1857. (App. V.2)

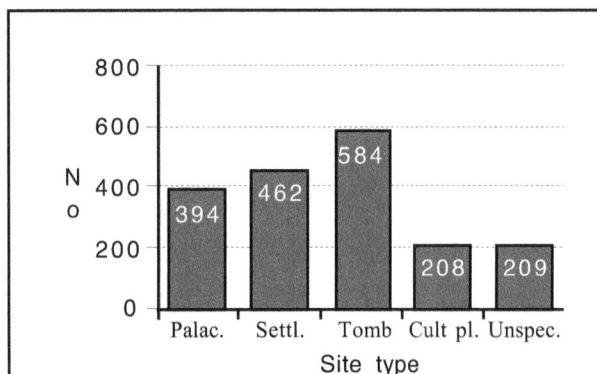

Fig. 6 Bronze objects by site type. Total 1857. (App. V.2)

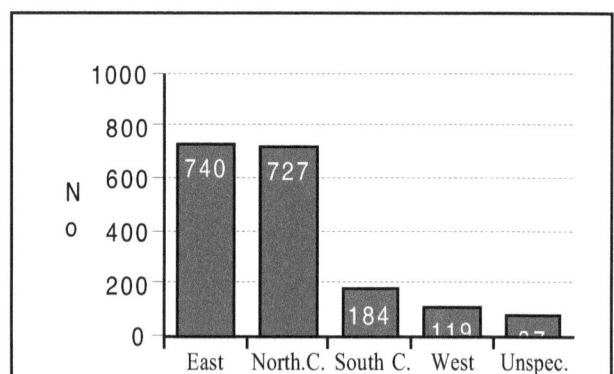

Fig. 7 Bronze objects by area. Total 1857. (App. V.2)

The tools dominate, a total of 451 objects or 53% of all (Fig. 8). The most common are chisels (79) and double-axes (77) followed by knives (42) and saws (39). A considerable number of the tools were found in Zakros (120), of which 60% came from the palace. The large amount of finds from Zakros can partly be explained by the fact that the palace was probably never robbed and partly by the detailed documentation of all tools in E. M. Platon's doctoral dissertation (1988). In Gournia, known as an industrial town, many tools have been found. In my opinion, perhaps the most interesting tools are the huge saws, the length of which could exceed 160 cm.

Many of the 112 bronze vessels are found in bronze collections in LM I destruction levels (Driessen and Macdonald 1997: 67-69). Georgiou prefers to talk about "metal groups" (1979). The word "hoard" is normally avoided in this context. The majority of the vessels were found in earlier excavations in Knossos, Malia and Gournia. The most impressive of them are perhaps the huge cauldrons from Tylissos, made of thick, riveted copper sheet. The weigth of the largest is 52 kg (Driessen and Macdonald 1997: 129).

The major part of the 119 weapons are daggers (77), mainly of a type developed from the Cretan Prepalatial long dagger. Of the 27 swords, the long MM III swords from Arkalochori constitute two thirds

of the total. Cult objects, or double-axes of sheet metal and cast figurines, are typical of the Neopalatial period. The database comprises 139 such objects, mainly small votive ones, but among them are also the four large double-axes from Nirou Chani and beautifully ornamented axes from the palaces, *e.g.*, the one from Zakros.

Of the bronze objects, 317 originate from palaces or their surroundings (Fig. 9). More than half of them come from villas or towns around the palaces, mainly Knossos and Zakros. All palaces had their own bronze workshops, but only slight evidence remains. More bronze objects are found in other towns and villas, totalling 350 objects or 41 %. Especially Gournia, Mochlos, Palaikastro and Hagia Triadha have yielded many fine bronzes. In the two first mentioned, convincing Neopalatial evidence for metallurgical activities has been found whereas the evidence from Palaikastro is dated to the Postpalatial period (App. I.2, I.3 and I.8). The bronzes from these sites were most probably made on-site. Neopalatial burials are extremely rare, and only 39 bronze objects have been found; some from tombs in the Knossos area and some from Mochlos. On the other hand, 103 objects come from cult places, mainly from Psychro and Arkalochori.

The regional distribution of the finds was as expected; 506 objects or almost 60 % of the total, were found in eastern Crete (Fig. 10). There were

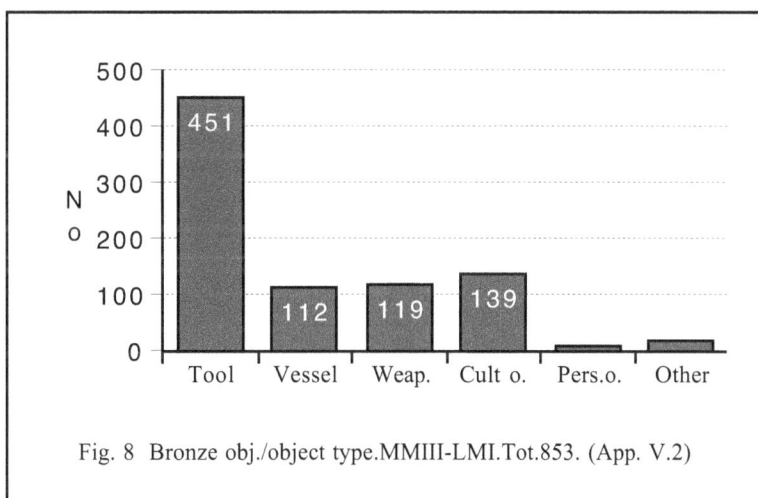

Fig. 8 Bronze obj./object type.MMIII-LMI.Tot.853. (App. V.2)

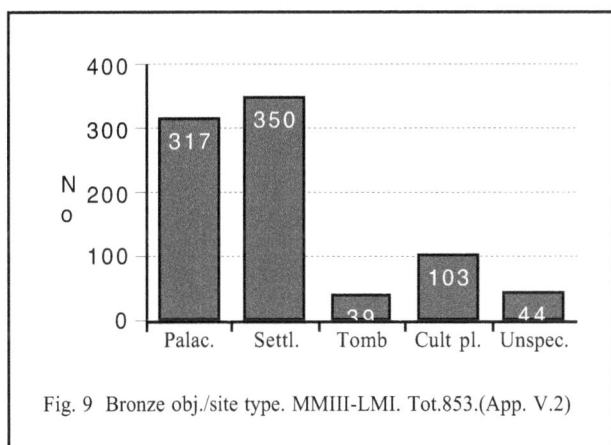

Fig. 9 Bronze obj./site type. MMIII-LMI. Tot.853.(App. V.2)

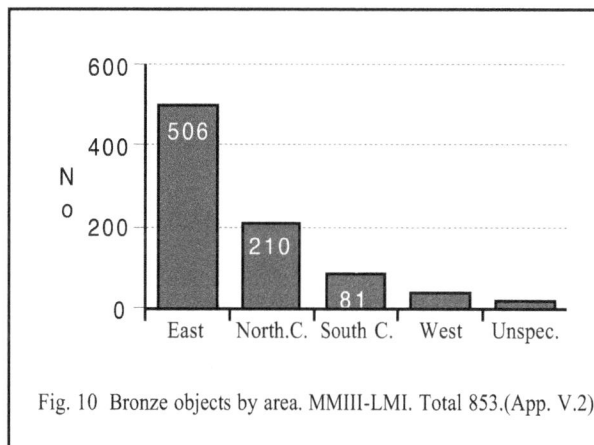

Fig. 10 Bronze objects by area. MMIII-LMI. Total 853.(App. V.2)

210 objects originating in north central Crete or the area around Knossos and Malia, while there are only 81 finds from south central Crete. Most of the 39 finds from western Crete come from Tylissos which, in accordance with the regions defined by Driessen and Macdonald (1997), belongs to the west.

The Mycenaean Knossos period (LM II - LM IIIA1)
The shares of the different object types, contexts and regional distribution for this period differ completely from the Neopalatial period. Tools no longer dominate, but weapons, mainly swords and spearheads are more numerous and more personal objects like razors and mirrors are found. This can to a great extent be explained by the fact that more than 80 % of the objects are from tombs in the Knossos area. But there might also be other explanations such as a new ethnic group, a different social system and new burial customs combined with possible changes in the availability of metals and new priorities for its use. (Figs. 11-13)

The 107 tools represent only 24 % of the total object assemblage. Half of them are knives, which perhaps better could have been classified as weapons because their length in many cases exceeds 30 cm. Tools are only on rare occasions found in burial contexts. The 72 vessels from this period were concentrated in a few graves which Catling and

Catling called "Burials with Bronzes". According to them, it was not a Minoan habit: it seems to have been introduced to Crete when other evidence indicates a "Mycenaean" presence. (Catling and Catling 1974: 253). The most important of these tombs are Sellopoulo, Tomb 4 (26 vessels), Zapfer Papoura, Tombs 14 and 36 (total 19 vessels) and Archanes, Tholos A (10 vessels). The 133 weapons, 30% of the total object assemblage, comprise 26 swords, 13 daggers, 32 spearheads and 61 arrowheads, 75 % of them originating from the tombs in the Knossos area. Other significant findspots are Chania, Pigi, Archanes and Armenoi. Notable is the low number of daggers, which during the Mycenaean Knossos period to a great extent were replaced by long knives. The arrowheads could, in many cases, have been used in hunting. The only Bronze Age bronze helmet from Crete, which was found in Tomb V at the New Hospital site, is included among the weapons. As many as 23 % of the finds were personal objects, mainly razors in men's graves but also ornaments and mirrors found in both men's and women's graves. (Fig. 11)

The assemblages of bronze objects from the Neopalatial and the Mycenaean Knossos periods are by no means commensurable. It is not only that the find contexts are different but the deliberately selected burial finds from the Mycenaean Knossos period give completely different opportunities for

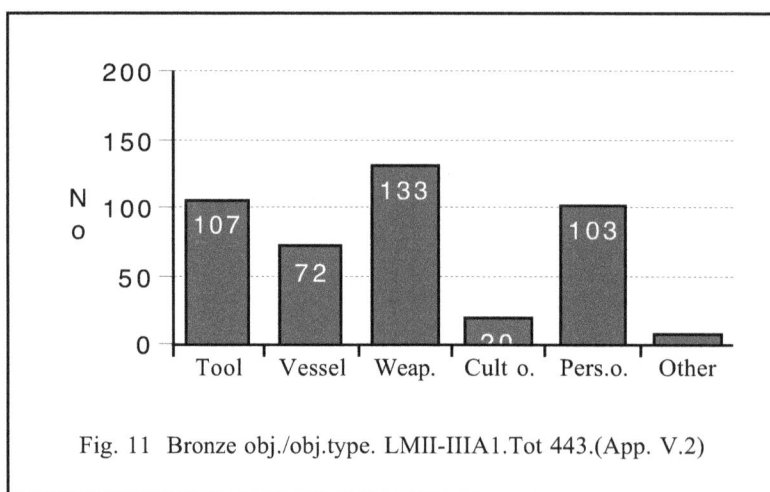

Fig. 11 Bronze obj./obj.type. LMII-IIIA1.Tot 443.(App. V.2)

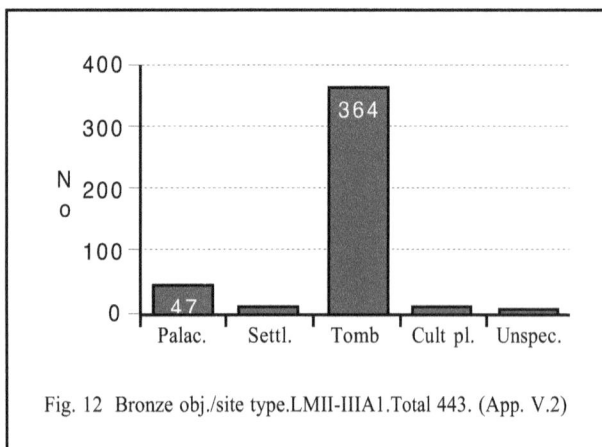

Fig. 12 Bronze obj./site type.LMII-IIIA1.Total 443. (App. V.2)

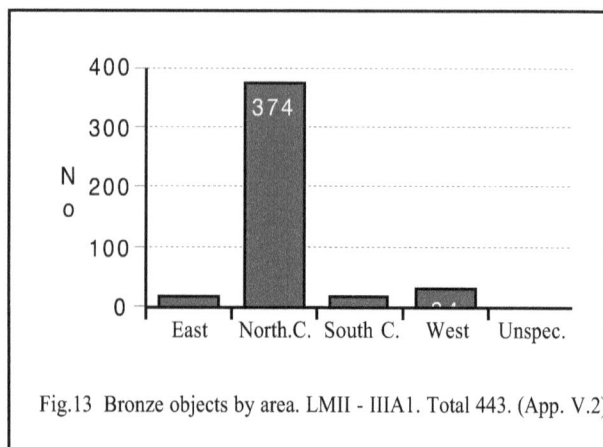

Fig.13 Bronze objects by area. LMII - IIIA1. Total 443. (App. V.2)

studying social structures and cultural habits than the randomly preserved objects from the Neopalatial period. For instance, the weapons and other burial finds from the "Warrior Graves" have always raised many speculations about who were actually buried in the tombs: Minoans or Mycenaeans? (*e.g.,* Popham 1974: 255-256; Niemeier 1983: 226; Matthäus 1983: 211; Kilian-Dirlmeier 1985: 209)

The Postpalatial period (LM IIIA2 - LM IIIC)

The database includes rather few finds, only 252, which can be securely dated to the Postpalatial period. The object assemblage for this period resembles that of the previous period, but differences occur (Figs. 14-16). The share of the tools increased from 24 to 32 % while for the vessels it decreased from 16 to 11 %. Almost half of the vessels come from the Mycenaean Burial Enclosure in Archanes. The share of weapons is almost the same, or one third, but the number of swords have increased from 26 to 42. The most important difference is that the finds from the Knossos graves no longer dominate so much. Burial finds still represent, however, the majority, but the findspots are spread all over the island. Among tombs outside the Knossos area are, *e.g.,* Mouliana and Myrsini in eastern Crete and Stamnoi in north central Crete. The finds from the large Postpalatial cemetery of Armenoi are so far not fully published. Settlement finds are rather few, only 28. Among the

finds from this period, the three Knossian swords in good condition found at Syme are worth mentioning.

The objects in the database from the long Postpalatial period are too few for an adequate comparison with the previous periods. They give only some general indications of the Postpalatial bronze industry.

4.1.3 Selected object types

Already by MM II, the Minoans had developed an involved series of metal tools, had produced the long sword and had developed a metal vessel industry, manufacturing vases with repoussé, that made their products influential even in the eastern Mediterranean (Betancourt 1998: 8). The Late Minoan bronze industry continued this development and some object types reached their acme during the Neopalatial period but in some cases later declined or disappeared from the archaeological record. During the Mycenaean Knossos period, new object types appeared, probably due to new ethnic groups and new customs.

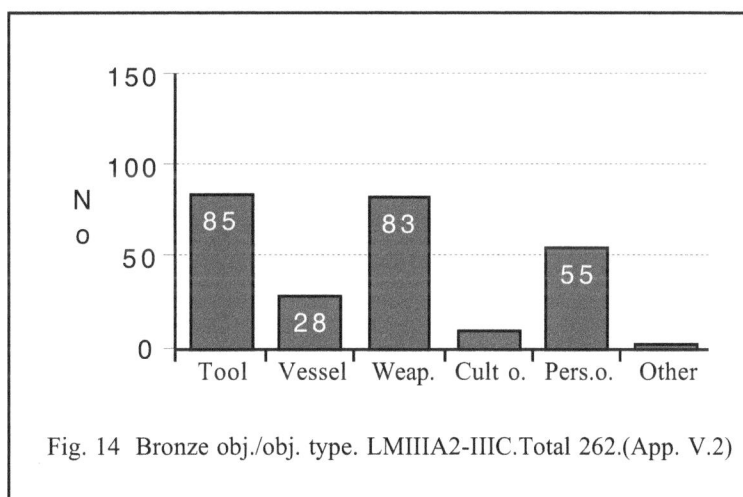

Fig. 14 Bronze obj./obj. type. LMIIIA2-IIIC.Total 262.(App. V.2)

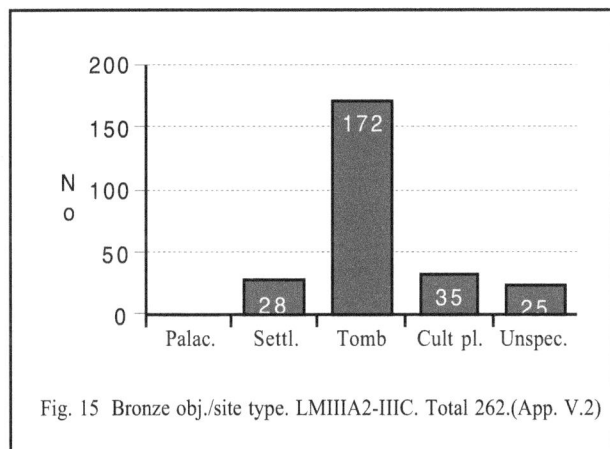

Fig. 15 Bronze obj./site type. LMIIIA2-IIIC. Total 262.(App. V.2)

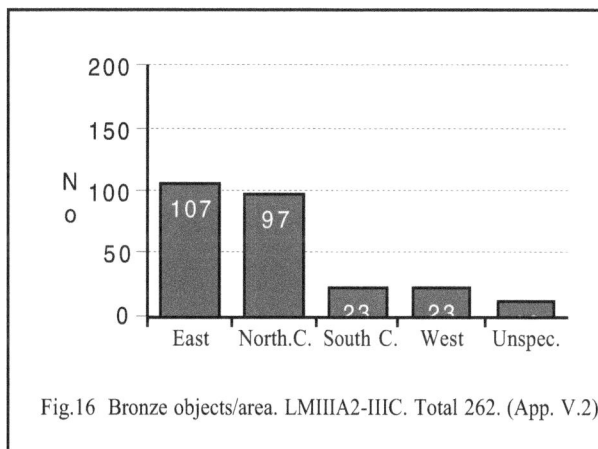

Fig.16 Bronze objects/area. LMIIIA2-IIIC. Total 262. (App. V.2)

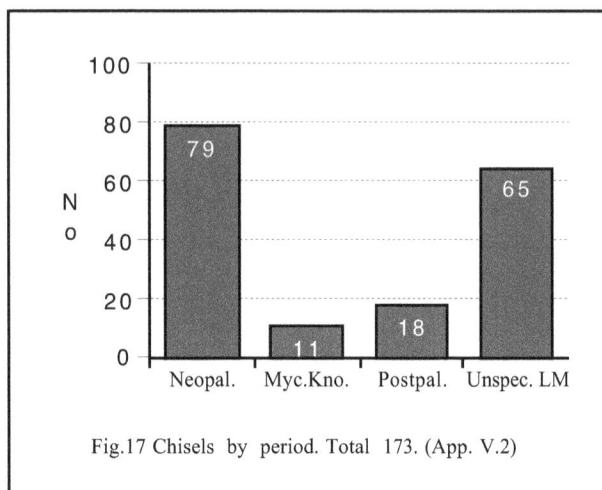

Fig.17 Chisels by period. Total 173. (App. V.2)

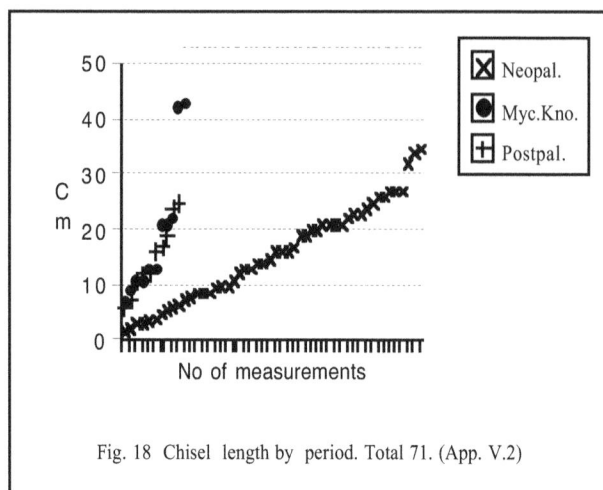

Fig. 18 Chisel length by period. Total 71. (App. V.2)

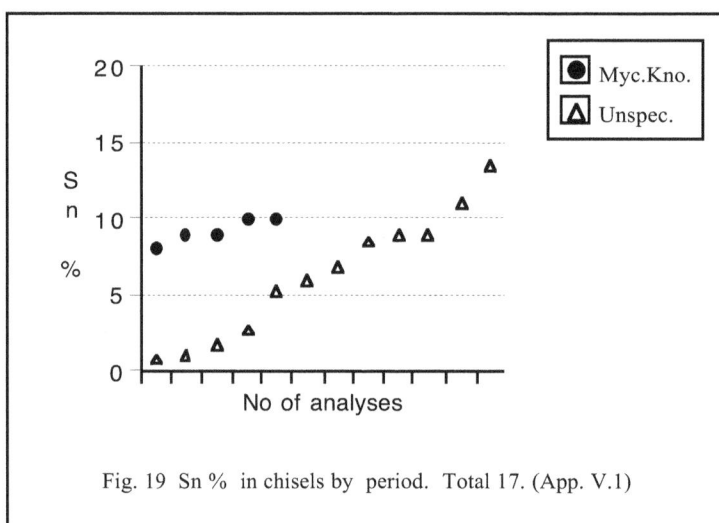

Fig. 19 Sn % in chisels by period. Total 17. (App. V.1)

In the following, the frequency of some selected object types in the database, their contexts, their characteristics and possible use are presented.

Chisels

Chisels are among the most numerous objects in the database, totalling 173 or 9% of all. About half of them, or 79, are dated to the Neopalatial period compared to only 11 chisels from the Mycenaean Knossos period, as tools were seldom deposited in graves. All the 11 chisels found are from the Un-explored Mansion at Knossos. During the Postpalatial period, burial customs seems to have been less strict; of the 18 chisels from this period, 12 were found in graves. A considerable proportion of all chisels, 38 %, cannot be securely dated, though probably most of them are Neopalatial. (Fig 17)

The size of the chisels varies from some few cm to 30 cm for heavy-duty chisels. The contexts where the chisels were found and the finds associated with them indicate that they were universal tools used by carpenters, masons, metalworkers and craftsmen doing small-scale work in other materials such as ivory. Loggers used them as wedges. Deshayes (1960) and Evely (1993: 2) have developed

typologies for the Bronze Age chisels based on physical criteria such as shape, size and robustness. These categories are, however, not used in this database. Only the dimensions and weights are stored, if available.

Chisels of all sizes are found in Neopalatial contexts. The few chisels from the Mycenaean Knossos and Postpalatial periods are in general larger, but due to so few finds, direct comparisons with the Neopalatial period cannot be made. (Fig. 18) Only 17 chemical analyses for chisels have been published. During the Neopalatial period, tin contents vary from 2.5 to 13% while the optimal level, around 10%, seems to have been the standard during the Mycenaean Knossos period (Fig. 19).

Double-axes

The massive, cast double-axes are typical of Crete. In the neighbouring areas, single-axes were preferred. Double-axes are the most numerous of the objects in the database, totalling 184 or 10% of all. About half of them are securely dated to the Neopalatial period while the rest have an unspecified dating. From the Mycenaean Knossos and the Postpalatial periods, there are hardly any double-axes at all

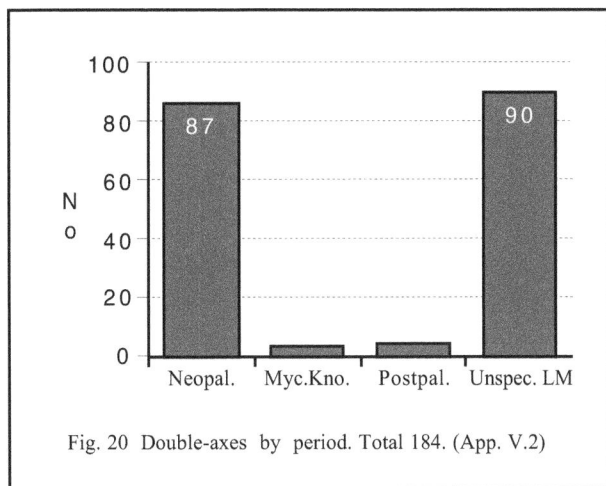

Fig. 20 Double-axes by period. Total 184. (App. V.2)

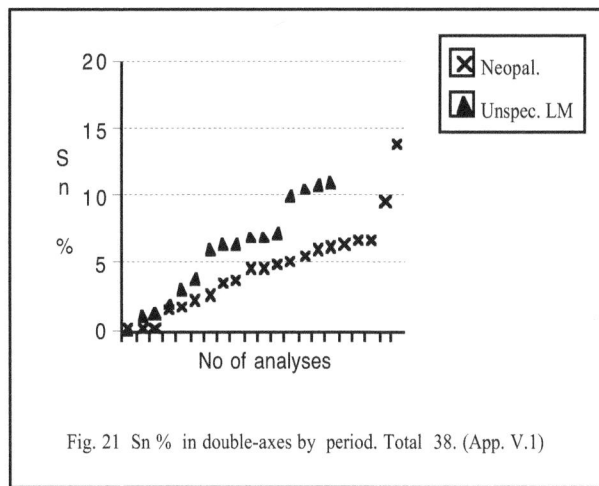

Fig. 21 Sn % in double-axes by period. Total 38. (App. V.1)

(Fig.20). The Neopalatial axes were found in palaces, particularly Malia (16 axes), Phaistos (10 axes) and Zakros (9 axes), and in the towns. The 90 axes with unspecified dates were found on more than 40 rural sites all around Crete. The lack of double-axes from the Mycenaean Knossos period was expected as tools normally are not found in graves. Perhaps more surprising is that double-axes were not found in Postpalatial contexts because the majority of LM III casting moulds are for double-axes (App. IV.2). According to Evely, the reason for the lack of axes from these periods might be increased recirculation of the metal rather than decreased use of the double-axe (Evely 1993: 55).

All the Cretan double-axes have almost the same shape. The ratio between length and width is rather constant, or 3:1. Of course differences occur and Buchholz has developed a typology for them (1959a: 8), which, however, is not used in this database. The length of the axes is between 16 and 20 cm, sometimes up to 25 cm, and the weight 0.9-1.4 kg. These heavy axes thus represented a fortune, especially for the rural population, and they were probable in use over a long period of time..

The tin content varies between 3 and 8% (Fig. 21). Some of the axes with unspecified dating contain more than 10% tin, indicating that they probably belong to later periods. Interestingly, some axes have a very low tin content, even less than 2.5%, which shows that the Minoans could achieve sufficient hardness for low-tin bronzes through repeated hammering and annealing. The axes often had to be sharpened and some finds indicate that worn-down axes were used as hammers.

Adzes
Some 20 adzes of different types have been found, among them double-adzes, axe-adzes and pick-adzes. The double-adzes are larger than the double-axes, their weight can exceed 1.5 kg. It is likely that they were agricultural tools, while the medium-sized axe-adzes and pick-adzes might have been used in woodworking, quarrying and dressing. The chemical analyses published show, surprisingly, that 6 double-

adzes out of 8 consist of almost pure copper. The rest, as well as the medium-sized axe-adzes and pick-adzes, have tin contents between 6.0 and 8.7%.

Saws
The Minoan craftsmen had a wide range of saws at their disposal; the database include 61 saws. Two thirds of them, or 39 saws, are securely dated to the Neopalatial period. There are 17 from Zakros and 9 from Gournia. Though 16 saws have unspecified dating only 6 saws have been found from the Mycenaean Knossos and Postpalatial periods. The majority of them are small. One exception is the well-known 48 cm long saw from Zapher Papoura, Tomb 33, which Evans called "The Carpenter's tomb". (Fig.22)

The Minoan saws can be grouped into three categories based on their length (Fig. 23): i) small saws (<10 cm) for precision work in precious metals, stone and ivory, ii) medium-sized saws (28-56 cm) for carpentry work and iii) large saws (>1 m) normally two-man implements for working timber. The largest saw in the database is 168 cm long and has a weight of 7.3 kg. These large saws are only found in palaces: 8 in Zakros and one each in Knossos, Malia and Hagia Triadha. This may indicate that tree felling for buildings and shipbuilding was a palatial privilege. These large saws are typical of Neopalatial Crete; they are not found elsewere. Later they disappeared. Can this perhaps indicate that shipbuilding on Crete ceased when the "Mycenaeans" arrived? Only medium-sized saws have been found in the towns, *e.g.*, in Gournia.

For saws, only four chemical analyses have been published. Surprisingly, they show that the large saws were made of almost pure copper while the small ones could have rather high tin contents. The reason for use of copper is probably that it is easier to hammer and cut out the teeth from a copper sheet than from the harder bronze. Catling and Jones were, however, surprised at these analyses, as they seems to be incompatible with the function of saws (1977: 65). But this is again an example of the metalworking skills of the Minoan bronze-

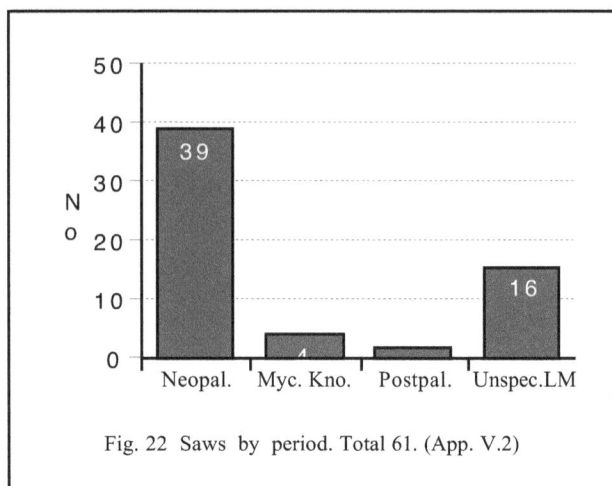

Fig. 22 Saws by period. Total 61. (App. V.2)

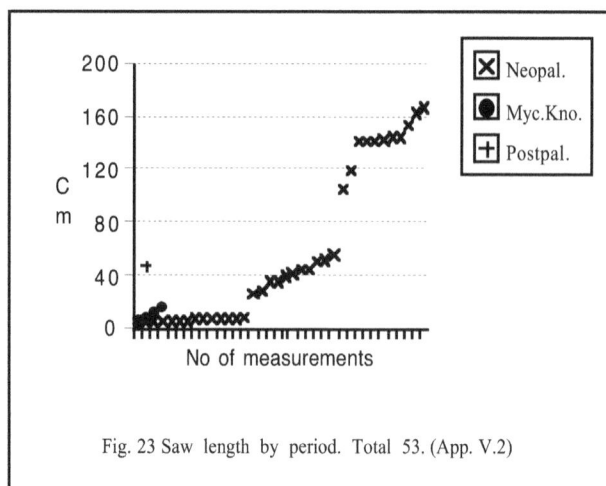

Fig. 23 Saw length by period. Total 53. (App. V.2)

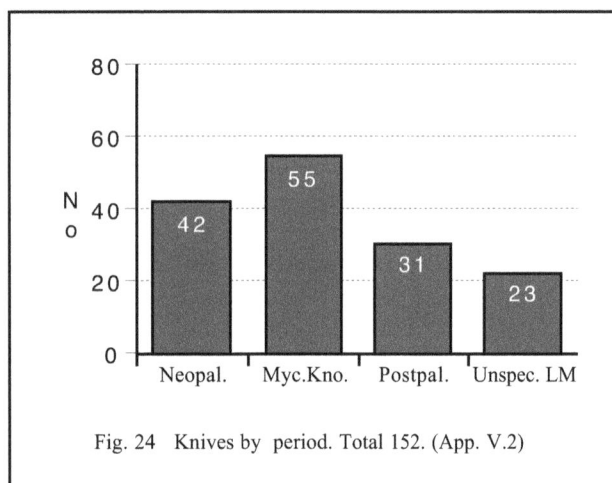

Fig. 24 Knives by period. Total 152. (App. V.2)

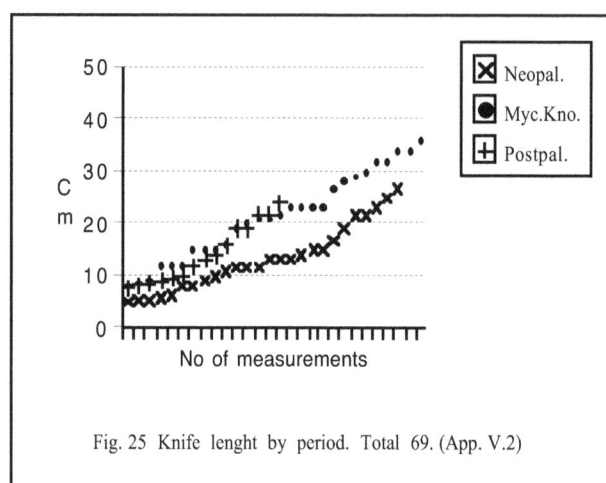

Fig. 25 Knife lenght by period. Total 69. (App. V.2)

smiths; they could achieve sufficient hardness by repeated hammering and annealing of pure copper sheets.

Knives

The small single-edged Minoan knife was suited to a variety of tasks: paring and whittling; chopping; sawing; scraping and cutting; carving; stabbing and piercing. Two-edged knives with long, pointed and symmetrical blades could perhaps better be classified as weapons. The database includes a total of 152 knives of both types, but the number might be too low; no extensive catalogue has been published for knives. Of this number, 42 or 28% are dated to the Neopalatial period and 55 or 36% to the Mycenaean Knossos period. This slight increase can be explained by the fact that daggers were to a great extent replaced by knives. During the Postpalatial period, the number of knives decreased, but not as drastically as other tools, giving an indication of the multi-uses of knives. (Fig. 24)

The Neopalatial knives were rather small; their length were normally less than 15 cm. During the Mycenaean Knossos period the so-called long-knives, found in the "Warrior Graves", were 20 - 35 cm (Fig. 25). There are 11 analyses for knives published, the tin content varying between 4.6 and 12% (App. V.1).

Vessels

The database includes totally 210 bronze vessels of which 102 or 48% are dated to the Neopalatial period (Fig 26). These were all found in settlements, except six small vessels from the Mochlos graves. A considerable number of them were found in bronze collections in LM I destruction levels, particularly in Knossos, Malia and Gournia. Perhaps the most impressive vessels are the huge cauldrons from Tylissos, made of thick riveted copper sheet, which are displayed in the Herakleion Museum. The weight of the largest is 52 kg (Driessen and Macdonald 1997: 129). The 72 vessels from the Mycenaean Knossos period were found in some few graves in the Knossos area. Large collections of bronze vessels were also found at Fourni in Archanes in the unplundered Tholos A, dated to LM IIIA1 (10 intact vessels) and in the Mycenaean Burial Enclosure, dated to LM IIIA2 (12 vessels).

Diachronic changes in the popularity of the vessel types are obvious; in general, the later vessels are smaller. For example the huge cauldrons of riveted copper sheet disappear after the Neopalatial period. The use of the different vessel types is not clear, and thus it is almost impossible to draw any conclusions of changing social habits from the vessel types.

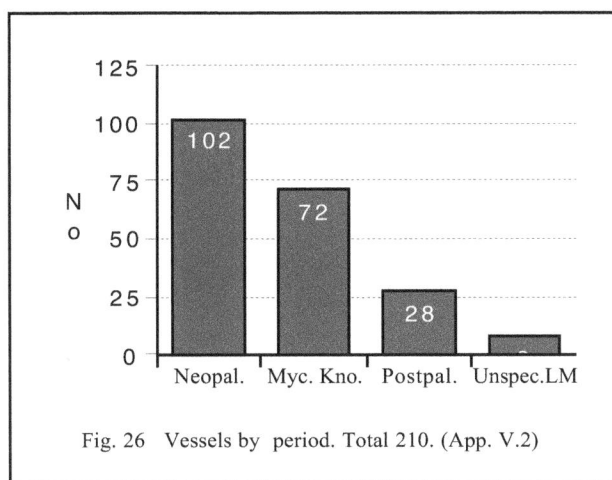

Fig. 26 Vessels by period. Total 210. (App. V.2)

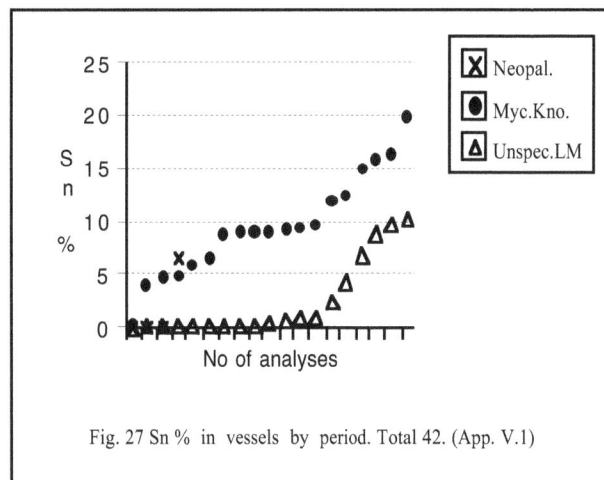

Fig. 27 Sn % in vessels by period. Total 42. (App. V.1)

Chemical analyses for 42 bronze vessels are published (Fig. 27). Of these, only one vessel is dated to the Neopalatial period and 22 to the Mycenaean Knossos period. Only unspecified LM is mentioned for the remainder. Their most common findspots, Knossos and Hagia Triadha, indicate, however, that the majority probably belong to the Neopalatial period. The analyses show that vessels of the Neopalatial period were likely manufactured by hammering almost pure copper sheets. In some cases, tin contents up to 10% have been found, but these analyses were made on the cast parts, *e.g.,* handles. From the Mycenaean Knossos period there is no evidence that vessels were manufactured from pure copper; the tin contents are mostly around 10%. It is difficult to explain why the use of pure copper ceased; one explanation could be that bronze is less toxic than copper.

Weapons
The database includes a total of 365 weapons: swords, daggers, spearheads and arrowheads as well as the only bronze helmet from Crete.

Of the 96 swords, 42 are surprisingly dated to the Postpalatial period, while the numbers from the Neopalatial and Mycenaean Knossos periods are lower, or 27 and 26 respectively (Fig. 28). During the Neopalatial period, the long swords from Arkalochori dominate (16 swords), but in both Malia and Zakros two long A-type swords have been found. All swords from the Mycenaean Knossos period were found in tombs except two. Normally there was only one sword in each grave. Exceptions are in Zapher Papoura, Tombs 36 and 44 which contained two swords each. Outside the Knossos area, swords from the Mycenaean Knossos period have been found in Archanes, Chania and Armenoi. The Ci- and Di- type swords dominate. They were probably all made in the Knossian workshop (Driessen and Macdonald 1984: 64). Of the 42 Postpalatial swords, 25 originate from graves all around Crete. They are of varying types and probably locally made. But the three beautiful swords found in Syme may be from the Knossian workshop.

The development towards shorter swords can clearly be seen in Figure 29. The length of the Arkalochori swords exceed one metre while the Neopalatial type A swords are 70-90 cm. The Ci- and Di-type swords are normally 40-70 cm. During the Postpalatial period, new types of shorter swords, the F-, Gii- and Naue-swords, were in simultaneous use with the Di- and Dii- swords. Normally, the optimal 10-12% tin bronze was used (Fig. 34). This may have been easier to achieve in the palatial weapon workshops which might have been priviledged and excluded the use of recycled metal (Baboula and Northover 1999: 151).

Of the 109 daggers of different types in the database, 77 or 71% are dated to the Neopalatial period (Fig. 30). During that period, the dagger was the main weapon, a status symbol and a universal tool. Of these, 32 come from Gournia and 18 from Malia. In the later periods, daggers were to a great extent replaced by long knives and spears as standard armament. The database includes only 13 daggers from the Mycenaean Knossos and 15 from the Postpalatial periods. The distinction of these later daggers, called dirks or short-swords, from the swords is solely based on their length. The boundary is, however, somewhat varying, in general 35 or 40 cm. During the Neopalatial period the daggers were short, between 10 and 30 cm. (Fig. 31). The Postpalatial Peschiera daggers were probably imported from Italy. Six such daggers have been found on Crete, five in the Psychro cave and one in Zapfer Papoura, Tomb 86 (Pålsson Hallager 1985: 295). Only a few analyses of daggers have been published (Fig. 34). Only the dagger from Gournia, studied by Charles, with a tin content of 13.1%, is relatively securely dated. It is one of the extremely few objects which has been metallographically investigated, generally referred to as "The First Sheffield Plate". (Charles 1968)

Of the total of 88 spearheads in the database, only 11 are dated to the Neopalatial period (Fig. 32). In Tomb XX on Mochlos, 3 intact spearheads were found, an indication that burials with weapons already occured among the Minoans. The rest are

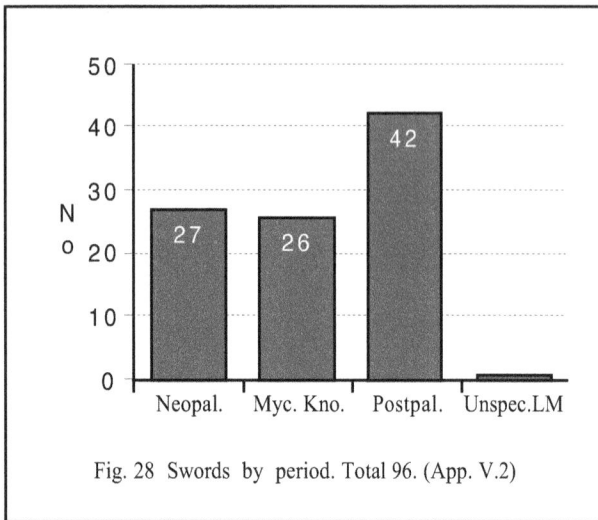

Fig. 28 Swords by period. Total 96. (App. V.2)

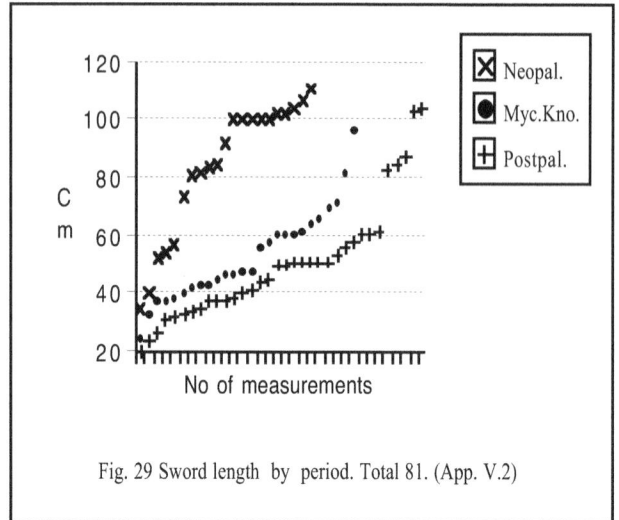

Fig. 29 Sword length by period. Total 81. (App. V.2)

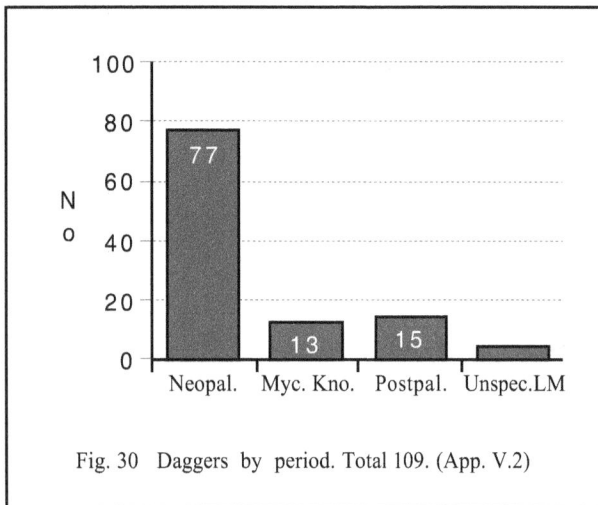

Fig. 30 Daggers by period. Total 109. (App. V.2)

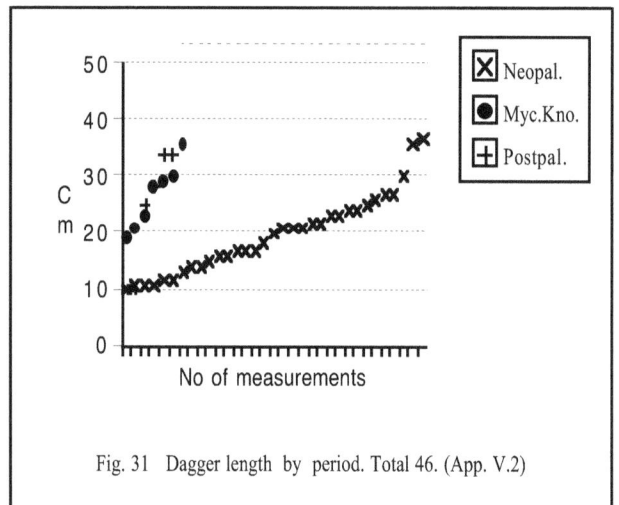

Fig. 31 Dagger length by period. Total 46. (App. V.2)

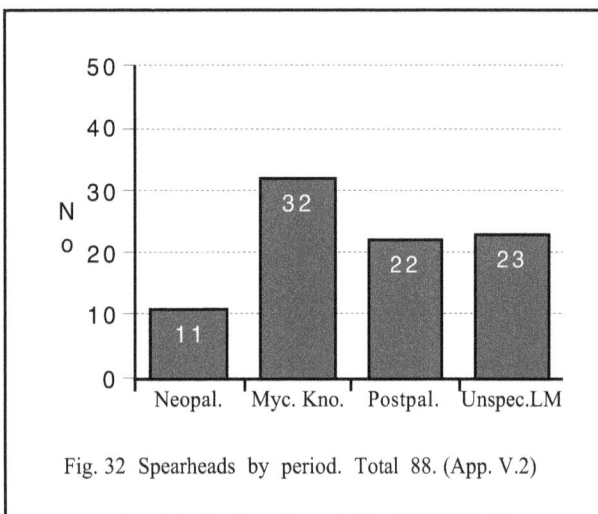

Fig. 32 Spearheads by period. Total 88. (App. V.2)

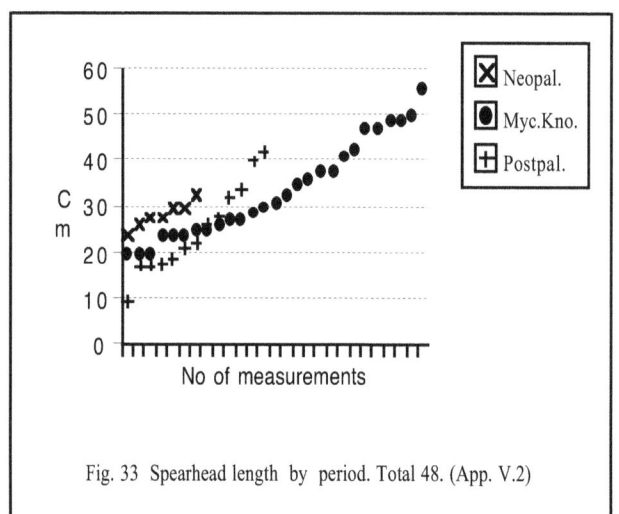

Fig. 33 Spearhead length by period. Total 48. (App. V.2)

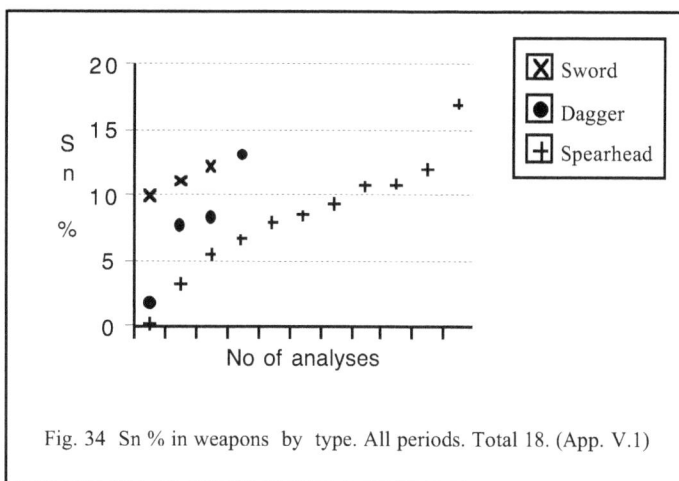

Fig. 34 Sn % in weapons by type. All periods. Total 18. (App. V.1)

from settlements: Gournia, Hagia Triadha, Malia and Pseira. During the Mycenaean Knossos period, the spear belonged to a warrior´s standard armament. In the Knossos area 29 spearheads have been found in tombs. Normally there was only one in each grave but Hagios Ioannis is an exception; Tomb 2 contained a whole armoury, including, *e.g.*, 6 spearheads and Tomb 1 had 3 spearheads. The 22 Postpalatial spearheads have been found all over Crete. Of the 23 spearheads with unspecified LM dating, the majority are from the Psychro cave.

The length of the spearheads varies considerably; during the Neopalatial period they were between 20 and 30 cm but during the Mycenaean Knossos period they could exceed 50 cm (Fig. 33). The 11 published analyses indicate a varying tin content The lowest values are from finds in the Ashmolean Museum, but the analyses are 50 years old and certainly no longer reliable. More recent analyses show an optimal tin content of 8 - 12%. (Fig. 34). The later spearheads were made by highly skilled craftsmen and are technically comparable to the swords. Probably they were also manufactured in the Knossian weapon workshop.

Cult objects

There are two types of bronze cult objects: cast anthropomorphic figurines and double-axes hammered from sheets of metal. Small axes of gold have also been found.

The database includes a total of 141 figurines, of which 95 are dated to the Neopalatial period, perhaps due to the fact that Sapouna-Sakellarakis´ study, from which the information is mainly collected, is concentrated on the Neopalatial period (Fig. 35). Male figurines represent 73% of the total. Their height varies between 5 and 25 cm, the female ones being somewhat smaller. The figurines were probably cast from scrap metal; pure tin was rarely added However, lead was added instead in many cases to improve the fluidity of the cast or to save bronze. As many as one third of the 51 analyses published show that lead was deliberately added; in some cases as much as 40% (Fig. 36).

Almost all published cult-axes of copper are dated to the Neopalatial period (Fig. 37). Of the 46 axes in the database, 26 come from Psychro. The majority of them are small, less than 10 cm wide. More spectacular are the large, ornamented, ceremonial axes found in cult rooms in palaces and towns, *e.g.*, the 47 cm wide axe from Zakros with double blades (N. Platon 1974: 143). Having special interest are also the four axes from Nirou Chani, more than 1 m wide, which probably were set on poles outside cult places.

Personal objects

Personal objects like razors, mirrors and ornaments are normally only found in tombs, and almost all

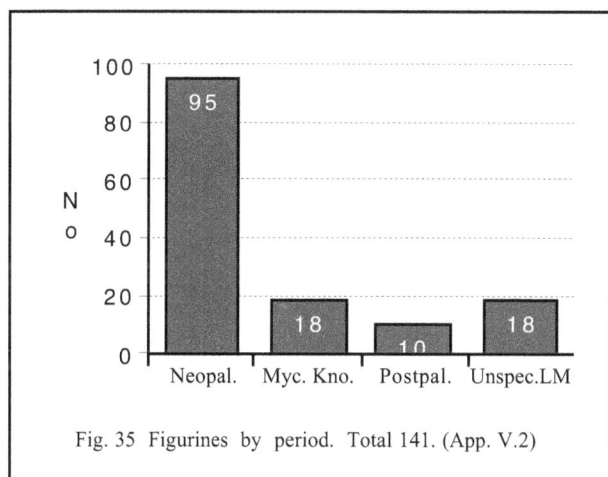

Fig. 35 Figurines by period. Total 141. (App. V.2)

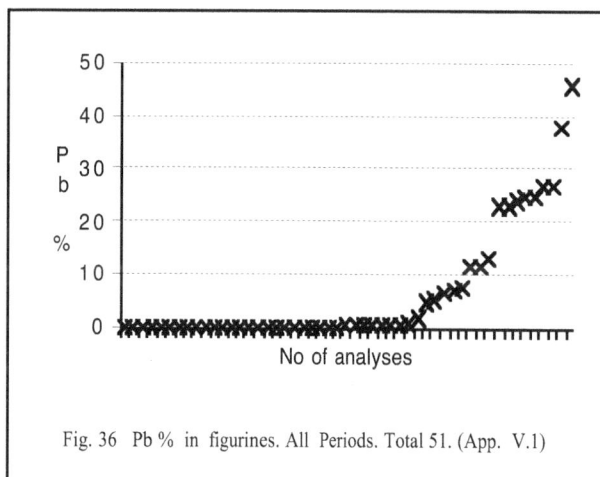

Fig. 36 Pb % in figurines. All Periods. Total 51. (App. V.1)

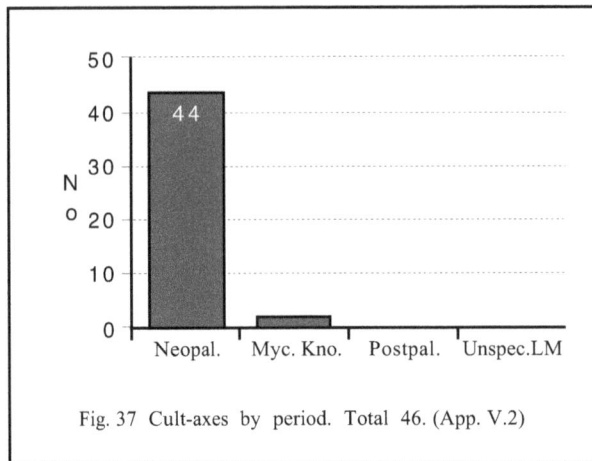

Fig. 37 Cult-axes by period. Total 46. (App. V.2)

objects in the database are thus dated to the Mycenaean Knossos and Postpalatial periods (Figs. 38, 40, 41). Razors and ornaments of bronze were already common in the Early Minoan period (Branigan 1968: 96-98) and the reasons for their being missing from the Neopalatial period can be the lack of tombs. But the Minoans probably never used mirrors; these might have come to Crete with the "Mycenaeans".

Of the 70 razors in the database, only four are dated to the Neopalatial period. The Minoans were seemingly clear-shaven, but they might have used ordinary knives and not the types found in the Knossos graves, which Evans called razors. Normally,

only one - but sometimes two - razors are found in men's graves. The length of the razors is rather constant, 15- 20 cm (Fig. 39).

All 43 mirrors are from tombs. Two thirds are dated to the Mycenaean Knossos, the rest to the Postpalatial period (Fig. 40). Normally only one mirror was found in each grave, but, *e.g.,* Sellopoulo, Tomb 4/I contained 3 mirrors and Tomb 4/III 2 mirrors. Their diameter varies between 10 and 15 cm; but a large mirror with a diameter of 22 cm was found in the Isopata Royal Tomb. All 58 ornaments in the database, except some rings from Mochlos, have been found in graves (Fig. 41). Considering the minimal metal use, the ornaments are of minor importance in this context.

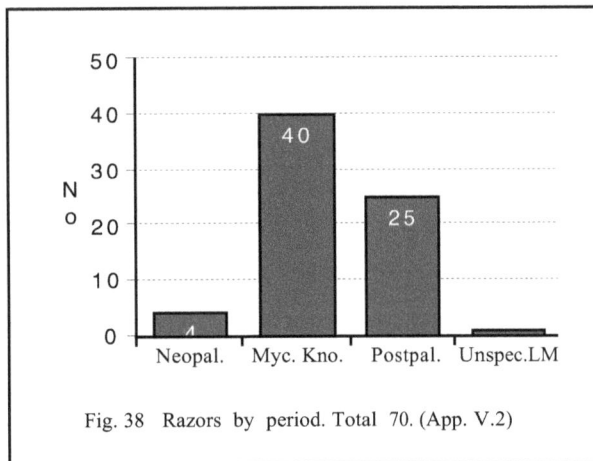

Fig. 38 Razors by period. Total 70. (App. V.2)

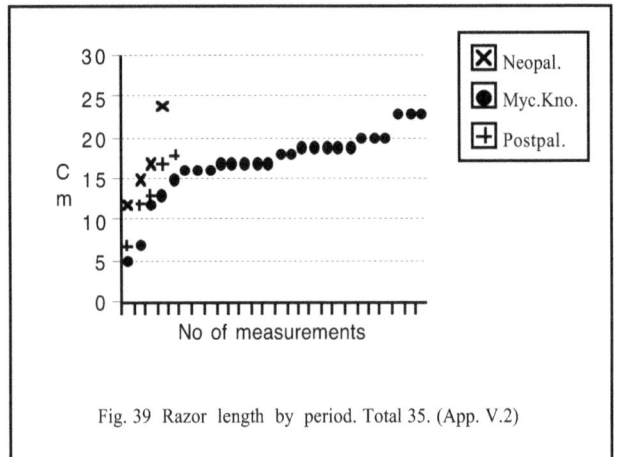

Fig. 39 Razor length by period. Total 35. (App. V.2)

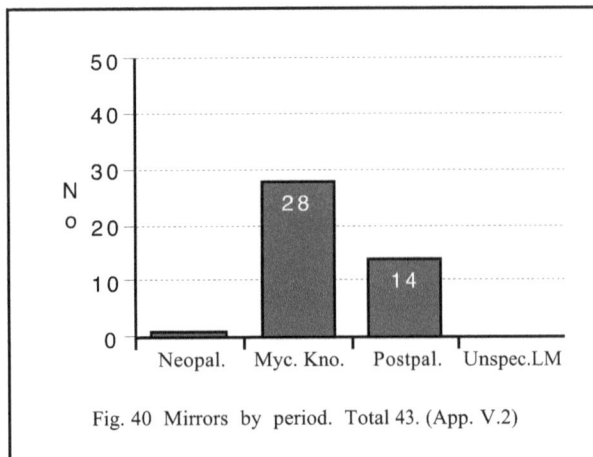

Fig. 40 Mirrors by period. Total 43. (App. V.2)

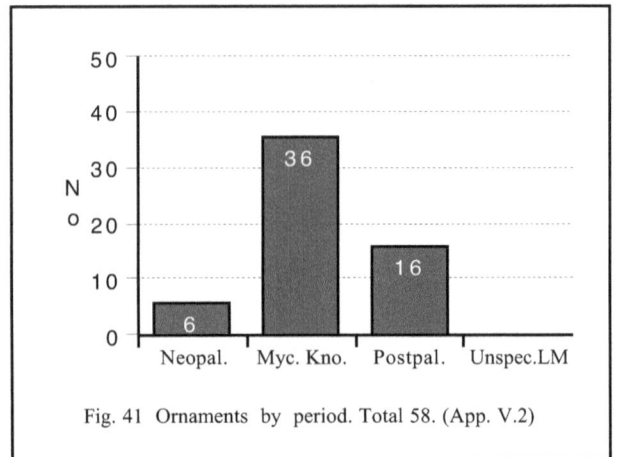

Fig. 41 Ornaments by period. Total 58. (App. V.2)

4.2 Bronze industry

4.2.1 Introduction

The interesting question of from where the Minoans imported their copper has been a central topic in Aegean archaeology over the last 20 years, partly due to the encouraging results achieved with lead isotope analyses. But the problem has mainly been studied in a broader perspective as a part of the international metals trade in the eastern Mediterranean and not specifically from the Cretan point of view. Despite important new finds, the problem, however, is far from solved. The question concerning the tin sources is even more complicated and will not be discussed in this study.

The archaeological evidence for bronzeworking and the analyses made on bronze objects are, so far, quite insufficient for an understanding of the bronze industry on Late Minoan Crete. But they give some indications concerning the raw materials used, the copper sources and the bronze technology. My intention is to analyze the data compiled in the databases and the conclusions which can be drawn from them. In the following, the raw materials for the bronze industry and possible copper ore sources will be discussed first: copper ingots and their provenience as well as re-melting of scrap metal. Then the evidence for the bronze industry is presented and the workshop equipment stored in the databases analyzed. Bronze technology is briefly discussed based on chemical analyses and metallographical studies of the bronze objects.

4.2.2 Raw materials

Alternative raw materials for the Minoan bronze industry were Cretan copper ores, imported ingots of copper, tin or bronze alloy and the re-melting of scrap metal. It has been a controversial question if the Cretan ores were used or not; so far no consensus has been achieved. Yannis Bassiakos from N.C.S.R. "Demokritos" in Athens believes that they were exploited at least to some extent, and he is presently involved in searching for evidence for Bronze Age copper mining and smelting (Personal comm. May 2003). I am, however, convinced that during the Late Minoan period the Cretan copper ores must have been quite insufficient for large-scale bronzeworking. The most probable raw material is imported copper ingots, either oxhide or bun ingots, both of which have been found on Crete. In contrast to mainland Greece, alloyed bronze ingots have not been found on Crete.

Copper ingots
A total of 30 intact and 39 fragments of copper oxhide ingots have been found on Crete. They are almost all dated to the Neopalatial period (App. III). Together they represent about one metric tonne of copper, a considerably amount of copper and a huge fortune. In comparison, all the 184 double-axes or alternatively the 96 swords in the database could have been manufactured from 175 kg copper. This indicates that the ingots were not only raw material for the bronze industry but, perhaps mainly valuable objects for storing or trading transactions. Only six intact bun ingots have been found, totalling about 30 kg of copper.

The findspots for the copper ingots are presented in Figure 42. Intact copper ingots have been find mainly on three sites: Hagia Triadha, Zakros and Tylissos. On the two first mentioned sites, the ingots were stored in the ceremonial area; in Hagia Triadha they were hidden in a cellar. Earlier they were supposed to primarily have been valuable objects, and called talents. It is difficult to understand why the ingots were stored in Hagia Triadha of the three main centres in Mesara. Phaistos, which according to the

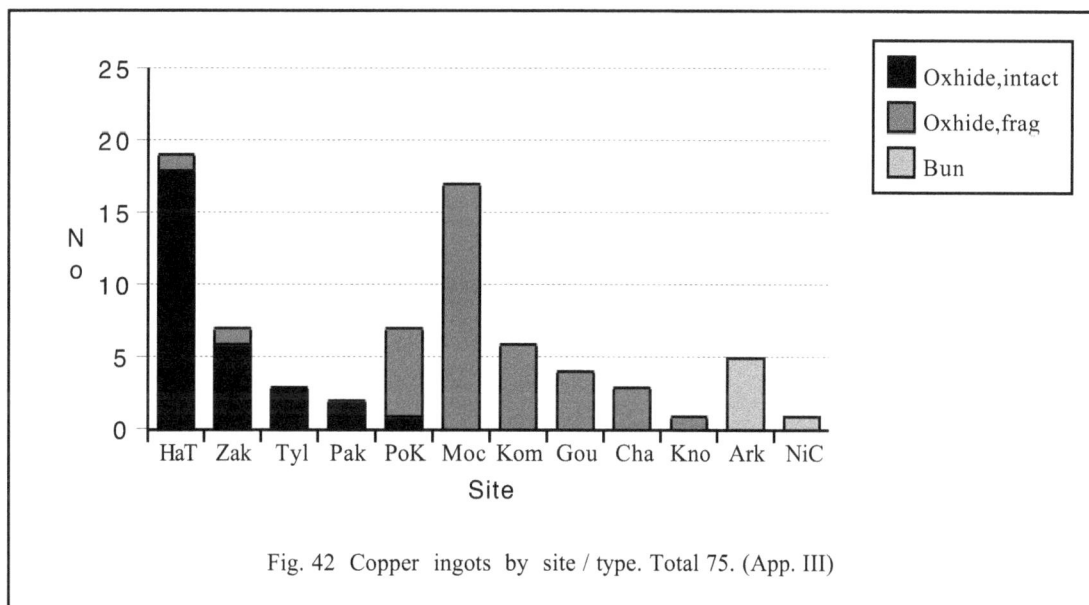

Fig. 42 Copper ingots by site / type. Total 75. (App. III)

myths was a prehistoric metalworking centre, seems to have been a more logical place. Even if it is now generally accepted that the remains of the large "metallurgical" kiln on the East Court most probably was a ceramic kiln (*e.g.,* Evely 2000: 301). Phaistos, like the other palaces, must have had a bronze workshop at least during the Protopalatial period. But so far no archaeological evidence for a workshop has been found. The ingots could perhaps have been tranferred to Hagia Triadha when Phaistos was destroyed or perhaps was it an intermediate storage for the ingots on their way from Kommos to Phaistos. Even if many bronze objects have been found in Hagia Triadha, is it difficult to imagine the "villa" as an industrial centre. No evidence for bronzeworking has been found there. The port of arrival for the ingots was probably Kommos, where evidence for rather large-scale metalworking during the Neopalatial period has been found (App. I.7). The six small fragments from Kommos are from a different type of ingot; either bun ingots (Blitzer 1995: 500) or Postpalatial oxhide ingots from Cyprus (Stos-Gale *et al.* 1997: 110).

Zakros was a most suitable place for storing the six intact oxhide ingots: a palace and a trading centre on the eastern coast with many finds of bronze objects and evidence for metalworking (App. I. 4). But the "metallurgical" kiln, published and reconstructed by N. Platon (1980), was in all probability not used for metal smelting (*e.g.,* E.M. Platon 1988: 187). In Tylissos, where three intact oxhide ingots were found, the situation is also not quite clear. The huge copper cauldrons, weighing altogether more than 100 kg corresponding to four oxhide ingots, must have been made on-site. But so far no traces of bronzeworking activities have been found.

The recent find of one intact oxhide ingot in Poros-Katsambas is one of the most important metallurgical finds since the 1960s. It is the first intact ingot found in a harbour town, in a context with evidence of bronzeworking (App. I.5). Earlier only fragments of ingots have been found in such bronze workshops and, it has been suggested that the palaces controlled the metal trade, distributing only ingot fragments to the workshops in the towns. Of course, Poros-Katsambas was the harbour of Knossos, and was probably controlled by the palace. The finds and metallurgical evidence are so far not completely published; only some brief articles are available (Dimopoulou 1997: 435). Additional fragments of oxhide ingots have been found on several sites in bronzeworking context, *e.g.,* in Mochlos, Kommos and Gournia (App. I. 2,3,7) and in Chania and Knossos.

Copper sources
A new era in the determination of the provenience for copper began when Noel Gale and Zofia Stos-Gale published their article on the applicability of lead isotope analyses also to copper (1982). For the last twenty years, the provenience of Late Minoan copper has been dealt with in numerous articles by the Oxford team (*e.g.,* Gale 1989, 1991, 1999; Gale and Stos-Gale 1986, 1987, 1992; Stos-Gale 1986; Stos-Gale and Gale 1994; Stos-Gale and Macdonald 1991; Stos-Gale *et al.* 1986, 1997, 2000). The results have been somewhat confusing and the problem with the copper sources for Late Minoan Crete is still far from solved. No summary or compilation of the results has been published. However, the results do indicate that the Minoans utilized many different copper sources and trading networks during the Late Bronze Age.

The provenience of the 19 ingots from Hagia Triadha, dated to LM IA, is a crucial problem. Lead isotope analyses indicate that they were made of copper from Precambrian ores which do not exist in the Mediterranean area. One possible source suggested is Afghanistan. Two of the Tylissos ingots seems to be made of the same copper as those from Hagia Triadha.. No secure source for the Zakros ingots has so far been published; most probably they came from Cyprus.

The ingot fragments from Gournia have been analyzed several times with varying results. The latest indicate Cyprus as the source (Stos-Gale *et al.* 2000: 209). The fragments from Mochlos are also reported to have been made of Cypriot copper (Soles and Davaras 2001). Three ingot fragments from Kastelli, Chania; one from Neopalatial and two from Mycenaean Knossos contexts, indicate Cyprus as the source (Stos-Gale *et al* 2000: 209). The Postpalatial fragments from Kommos most probably also came from Cyprus (Stos-Gale *et al.* 1997). For the Poros-Katsambas ingot and ingot fragments, no analyses have yet been published. The bun ingots were likely to have been made of copper from Laurion, but no lead isotope analyses have been published.

Lead isotope analyses for the copper in bronze objects have further complicated the situation. No reliable compilation of all analyses on firm dated objects have been published. Only some few stray analyses on often not clearly specified objects can be found in different articles. These indicate that, during the Neopalatial period, the main source for the copper in bronze objects was Laurion, but copper from Cyprus also occurs, particularly in eastern Crete. During the Mycenaean Knossos and Postpalatial periods, copper from Cyprus probably dominated. An enigma is why the copper from an unidentified source in the Hagia Triadha ingots has not have been found in any object. Were these ingots not used at all as raw material for bronzeworking?

Scrap metal
Scrap metal has been found together with fragments of copper ingots in several bronze workshops indicating that it was used as a supplement to the expensive and problematically available virgin metal. The best evidence is from the thoroughly investigated bronze workshop in the Unexplored

Mansion at Knossos, dated to LM II, where, however, no ingot fragments were found (Catling and Catling 1984). In Mochlos, foundry hoards of broken bronze vessels together with ingot fragments have been found (Soles and Davaras 1996: 201). It is impossible to estimate the role and share of scrap metal in the Minoan bronze industry. A crucial factor must have been the availability of virgin metal. Optimal tin contents in, *e.g.,* weapons indicate that the use of metal might have been controlled. The virgin metal could have been reserved for palatial weapon workshops.

4.2.3 Bronze workshops

Based on the numerous bronze objects and copper ingots, one could anticipate that there were bronze workshops in every palace, town and larger villa and even in the countryside. The bronze object assemblage comprised not merely prestige objects for an elite, but to a great extent also tools for ordinary people. However, the archaeological evidence for bronze workshops is surprisingly scarce. We have evidence only for some ten workshops, all periods included. A reason might be that metallurgical evidence was not recognized and recorded in older excavations. Fortunately the situation has changed, and crucibles, moulds and metallurgical waste have been found in many new large excavations. The workshop equipment found are presented by workshop in Figure 43, and the main workshops briefly outlined in Appendix I.

From the Protopalatial period, we have good evidence for a workshop in Quartier Mu, Malia, but the evidence for palatial workshops from the Neopalatial period, which certanly existed, is scanty. Only one crucible and one mould have been found at Zakros; the "metallurgical" kiln was probably a pottery kiln. From Malia, we have the 12 well-

preserved stone moulds, probably from a stone workshop, but the only sign of metallurgical activities is a burnt spot. There are a few crucible fragments found in different rooms at Knossos. In Phaistos the only clay mould for lost-wax casting found was for a fist of natural size, indicating technologically advanced, large-scale bronzeworking already during the Protopalatial period. The workshops were in general located outside the palaces, probably for health reasons.

The best evidence for bronzeworking is from coastal towns with direct access to imported metal. Gournia, where Boyd Hawes found five open stone moulds, has always been considered as the industrial town *par exellence.* Additional evidence consists of four copper ingot fragments, scrap metal, slag and tools probable used in metalworking. (App.I.2) Recent excavations in Kommos, Mochlos and Poros-Katsambas have revealed convincing proof of metallurgical activities. During the Neopalatial period in Kommos, large-scale metalworking activity was concentrated to the public buildings in the southern area. Evidence includes crucible fragments, metal bars, wire and strips intended for manufacturing, slag, debris and tools. In LM II-LM III metalworking had apparently ceased in the southern area and continued as small-scale operations in the settlement areas on the Hilltop and Central Hillside. The abundant evidence includes, *e.g.,* crucibles of a different type, clay investment moulds, pot bellows, ingot fragments and slag. (Blitzer 1995). (App. I.7) Kommos, with metallurgical evidence from two different periods, offers the best opportunity for studying the development of bronzeworking during the Late Minoan period. In Poros-Katsambas recently numerous signs of metal-working have been found, dated from MM II to LM IIIA2-LM IIIB, but mainly from the Neopalatial period. Among the finds are an intact copper oxhide ingot, a lead ingot, fragments of crucibles, moulds, tuyéres and pot bellows as well as slag, metal scrap and a crucible

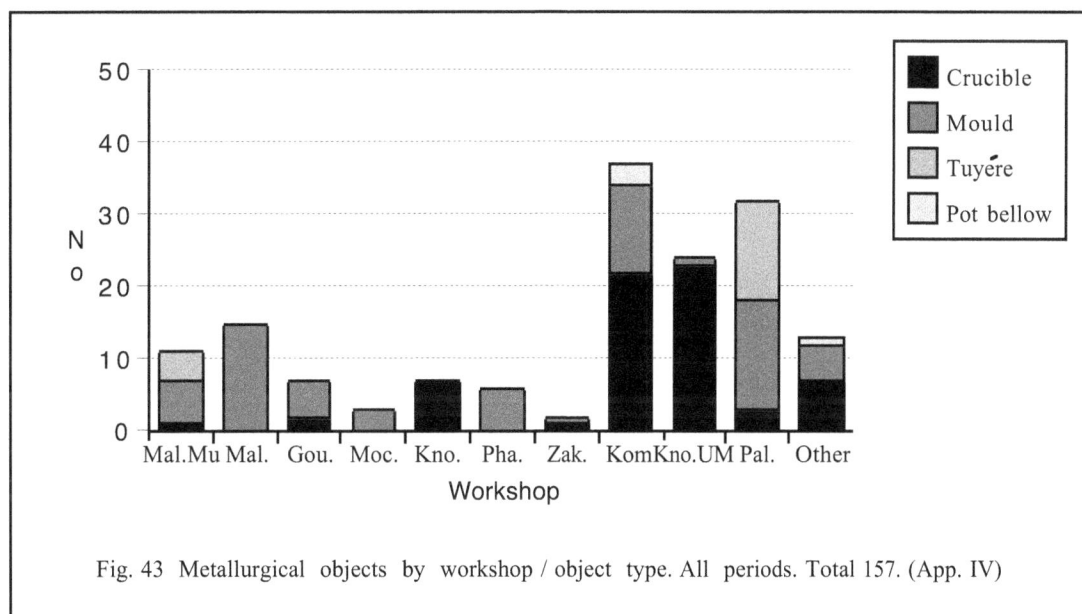

Fig. 43 Metallurgical objects by workshop / object type. All periods. Total 157. (App. IV)

furnace as well as a constructed rectangular furnace (Dimopoulou 1997). (App.I.5) No detailed information, however, has so far been published and the finds are thus not included in the databases. At Mochlos, on the coast opposite the island, LM IB workshops for, *e.g.,* bronzeworking have been found (Soles and Davaras 1994). (App. I.3)

By far the most thoroughly investigated Late Minoan bronzesmith´s workshop is in the Unexplored Mansion at Knossos with over 300 objects related to metallurgical activities scattered around in several rooms. The actual work-area may have been on the upper floor. Its equipment included some small furnaces, a brazier for transporting lighted charcoal and several crucibles. But only one small mould was found. The tool-kit included pincers, drills and awls; of the products, mainly basic casts as bars and billets have been found. No virgin metal was recovered, but there was a large quantity of scrap metal apparently for re-melting. The workshop is dated to LM II (Catling and Catling 1984). (App. I.6)

There is so far little evidence for Postpalatial bronzeworking. The best comes from Kommos (see above) but a pit with metallurgical debris dated to LM IIIB was recently investigated in Palaikastro. The finds comprised mainly fragments of crucibles, moulds and tuyéres.. The actual workshop area, however, has not been found. (Hemingway 1996). (App. I.8)

4.2.4 Metallurgical objects

In total fragments of only 157 workshop equipment have been found: 66 crucibles, 69 moulds, 18 tuyéres and 4 pot bellows from 12 sites. The finds from the Protopalatial Quartier Mu in Malia and the few Prepalatial finds are also included. The databases are presented in Appendix IV and the distribution of the finds by workshop and equipment type in Figure 43.

The most numerous finds, 37 in all, are from Kommos, where evidence for metalworking has been found from both the Neopalatial and the Postpalatial periods. The 32 finds from Palaikastro, are from a Postpalatial debris pit. From the LM II workshop in the Unexplored Mansion at Knossos, with more than 300 metallurgical objects, only 24 implements of the types in Appendix IV are found. It is notable that among them is only one mould. From Malia and Gournia we have the famous stone moulds already found hundred years ago. In the recently excavated Protopalatial Quartier Mu in Malia, fragments of four tuyéres have been found, which might be the oldest in the Aegean (Poursat 1996: 117).

The workshop equipment clearly indicates a change in bronze technology after the Neopalatial period.

New types of both crucibles and moulds appear in the archaeological record during the later periods (Figs. 44, 45). The Neopalatial crucibles were massive, deep and thick-walled clay bowls mounted on a substantial pedestal base, which was perforated for the wooden rod for lifting the crucible. They reflected a metal-melting tradition which first appeared on Crete in EM I. The database comprises 17 crucibles of this type, mainly from Kommos. They were replaced by smaller, hemispherical, bridge-spouted crucibles of a different fabric. These crucibles were probable lifted with tongs. The database comprises 38 crucibles of this type from the Unexplored Mansion in Knossos, Kommos and Palaikastro.

The sharp chronological division between the large Neopalatial and the later hemispherical bridge-spouted crucibles is apparent in Fig. 44. But this might be a simplification based only on finds from some few sites. The causes for this change cannot be the arrival of the "Mycenaeans", as hemispherical spouted crucibles have been found in much earlier contexts, *e.g.,* the small Protopalatial crucible from Quartier Mu (Poursat 1996: 116-117, Pl.55) and the unpublished Neopalatial crucibles from Palaikastro (Evely 2000: 349). But at any rate some change occurred, and the seemingly impractical crucibles on a high pedestal are no longer found after the Neopalatial period.

Of the total of 36 one-piece open or two-piece stone moulds intended for mass production, preferably of schist or limestone, 27 are dated to the Neopalatial period, the remainder to the Protopalatial period. These moulds were valuable, as attested by their heavy wear and assiduous repair. By far the most common object cast in the two-piece moulds is the double-axe; the open moulds were used for flat objects such as chisels, bars and knives. The stone mould, however, almost disappeared after the Neopalatial period and was, based on the few available finds, replaced by clay investment, single-cast moulds for lost-wax casting. (Fig. 45) From the Postpalatial period, there are 25 such moulds from Kommos and Palaikastro. They consist of two clay layers of different fabric enwrapping a wax core. All 12 moulds from Kommos were for the production of double-axes with shaft holes and the 13 from Palaikastro for, *e.g.,* double-axes, other axes, bars and decorative elements.

The reasons for the changes in casting technique are not known. So far, only guesswork has been presented. Concerning the changeover from stone moulds to clay investment moulds, Hemingway suggests that the impermanence of the clay moulds better suited the smaller scale LM III metalwork than the stone moulds, intended for mass production. An additional reason might have been the apparent discontinuity of specialized stonecarving workshops (Hemingway 1996: 241). But Catling states that the evidence so far is quite insufficient to draw such a

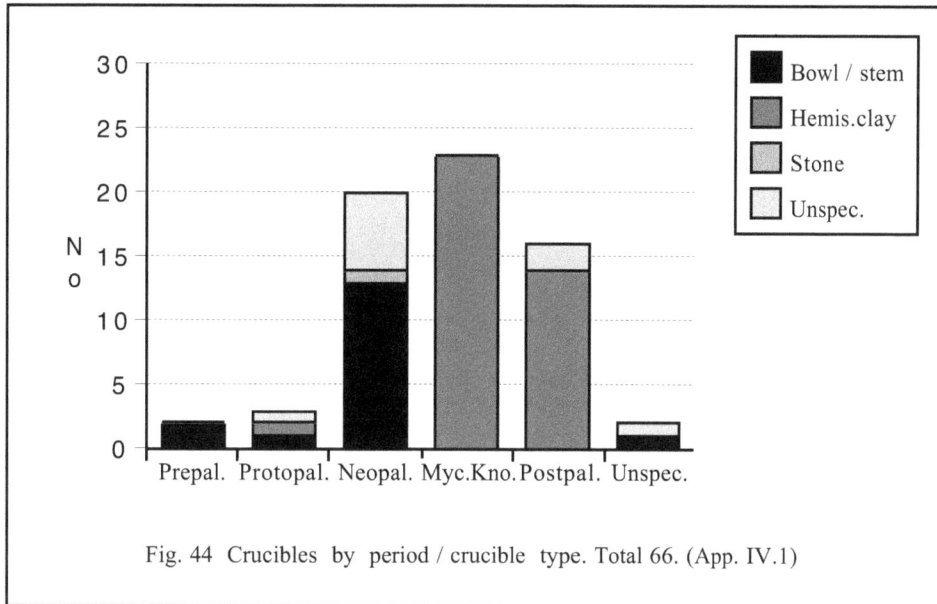

Fig. 44 Crucibles by period / crucible type. Total 66. (App. IV.1)

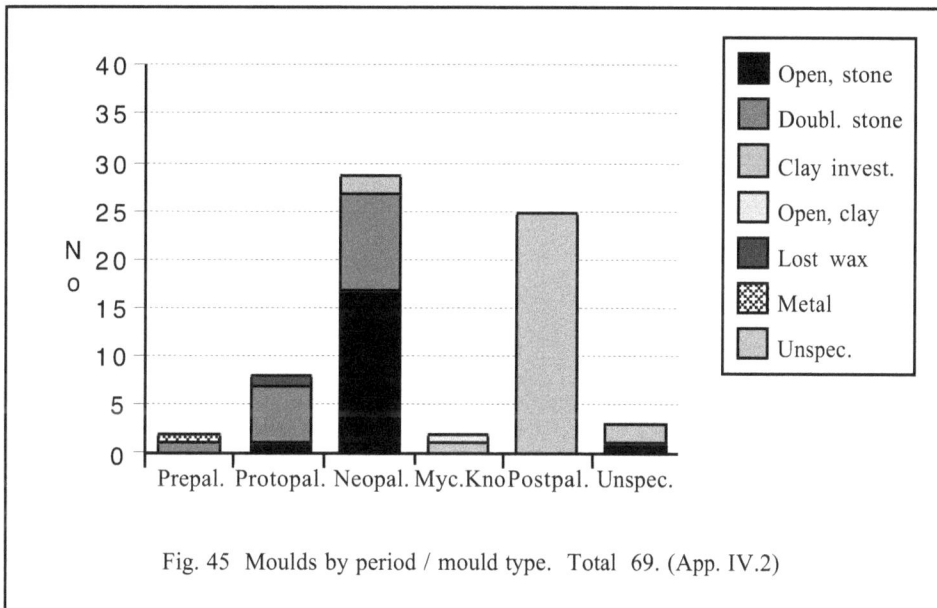

Fig. 45 Moulds by period / mould type. Total 69. (App. IV.2)

contrast between Neopalatial and Postpalatial metal-working and reminds us that the evidence from Palaikastro is from a rubbish pit whereas a stone mould was of lasting value (Catling 1997: 56). The stone mould from Pseira, dated to LM IIIA2 - LM IIIB, is an indication on that stone moulds might have been still in use during the Postpalatial period.

So far, very few scientific investigations of ancient "metallurgical ceramics" have been done. Recently, however, the composite material for the crucibles and moulds from Kommos has been investigated by C. Oberweiler from the University of Nanterre-Cedex, France in cooperation with N. C.S.R. "Demokritos", Athens. The results show that the wares used by the Minoans during the Neopalatial and the "Mycenaeans" during the Postpalatial periods have practically the same properties, or increased thermal shock resistance. But the manufacturing process and the raw material used were quite different. The Minoan bronzesmith made use of the extended microporosity obtained naturally by calcareous clay, while the "Mycenaean" bronzesmith improved the properties of the non-calcareous clay they used by adding a mixture of quartz temper and vegetable fibres. The reasons for this difference are not clear, but it can be the effect of both the traditional craftmanship know-how of each people and a technically concious development process. The project will continue by analyzing material from other Bronze Age sites on Crete. (Oberweiler *et al.* 2003)

Chapter 5. Development of Late Minoan bronzeworking

5.1 Introduction

The division of the compiled Late Minoan bronze objects and working equipment into the Neopalatial, the Mycenaean Knossos and the Postpalatial periods gives some indications of the development of Late Minoan bronzeworking. Further analyzing the data for each period forms the basis for recognizing the development pattern, stating some preliminary salient features of the bronzeworking for the different periods and identification of the most apparent changes and trends. More important are the difficult questions of the reasons for and consequences of this development, and even more problematic, the impact of metals on Minoan society. Many more multidisciplinary investigations are, however, needed for finding answers to these questions. This study is mere a presentation of the general pattern.

5.2 The Neopalatial period (MM III - LM IB)

It is self-evident that bronze and skilled bronze-smiths were essential for a palatial Bronze Age economy. Crucial uses of the metal were tools for building of palaces and ships, weapons for defence and as prestige objects for the elite and for exchange. During the Protopalatial period, the situation was almost the same, but the evidence is considerably less, for instance, the famous prestige swords from Malia, the abundance of bronzes in the Arkalochori cave, the sophisticated lost-wax mould from Phaistos and the bronze workshop in Quartier Mu in Malia. Bronze can even have been the impetus for the emergence of Minoan foreign contacts with the Near East during the Middle Minoan period (*e.g.,*Wiener 1990; Betancourt 1998; Hitchcock 2001).

During the Neopalatial period, the acme of the palatial culture, the demand for bronze objects increased. Effective tools were needed for the massive, monumental building programmes, the elite were accumulating prestige objects such as weapons and bronze vessels and the cult objects of clay were replaced by bronze. So far, about 850 Neopalatial bronze objects have been published, mainly settlement finds from eastern Crete, the Knossos area and Mesara. The distribution of the finds indicate that western Crete was only sparsely inhabited, but the reason for this can also be the focus of archaeological activities. The majority of the finds are tools: double-axes and chisels were all-round tools used by everybody, huge two-man saw of hammered sheets of copper were used for tree-felling, probable under palatial control, and small tools needed for precision work. The Minoans are supposed to have been peaceful and relatively few weapons have been found. The majority of them were daggers, in accordance with the old Minoan habit of wearing a dagger in the belt. The swords were long and cumbersome and probable never intented for warfare. Some of them were votive-swords. The religious ceremonies were concentrated in peak sanctuaries and sacred caves where hundreds of small cult-axes of hammered gold-, silver- or bronze sheet metal and cast bronze figurines have been found.

Only tin bronze and pure copper were used; the arsenical bronzes no longer were made. All copper and tin for the large-scale bronze industry had to be imported. The earliest oxhide ingots of copper in the Aegean area are found on Crete in Neopalatial contexts. The identification of their sources by lead isotope analyses has so far been unsuccessful. During the end of the period, copper ingots from Cyprus also appeared on Crete. The main copper source during the Neopalatial period was, however, probably Laurion. The tin sources for the Aegean area remain an enigma. On Crete neither tin ingots nor ingots of tin bronze have been found. Particularly in workshops outside the palaces, re-melted scrap metal was also used. .

The bronzesmiths were highly skilled and had fully mastered bronze metallurgy. Few entirely new innovations were made, but the technique, principally used since the Early Minoan period, was continuously developed. The clay crucibles used reflect a particular Aegean metal-melting tradition, known on Crete since the Early Minoan period. The one- or two-piece stone moulds were apparently intended for massproduction.

The Neopalatial period ended with a series of catastrophes triggered by the Thera eruption which bought the Minoan empire to the edge of a ruin. The culmination was the LM IB destruction of almost all settlements except Knossos, most probably by human action (*e.g.,* Driessen and Macdonald 1997). Important indications of the social unrest are the collections of bronzes and the isolated or hidden weapons found in LM IB destruction levels.

5.3 The Mycenaean Knossos period (LM II - LM IIIA1)

The Minoan administration almost collapsed in LM IB and Crete was partly depopulated. The "Mycenaeans" or "Mycenaeanized" Minoans seemingly arrived in some kind of a political and administrative vacuum. With them, a more warlike society with centralized administration was introduced in Knossos with new customs such as burial habits, attested, *e.g.*, in the "Warrior Graves", and new cults. The peak sanctuaries and sacred caves lost their importance.

These changes are reflected in bronzeworking as a changed structure of the bronze object assemblage and the find contexts. Tools no longer dominate, the proportion of weapons increases and almost all 443 bronze objects from this period are found in tombs. The few settlement finds almost all originate from the Unexplored Mansion at Knossos. This indicates that the "Mycenaean" control was probably limited to central Crete. Bronze objects in men's graves seems to have been rather standardized: some weapons, *e.g.*, a sword, a short-sword, a spearhead, a knife and some personal objects, *e.g.*, a razor, a mirror and ornaments. In a few graves, "The Burials with Bronzes", large collections of bronze vessels were found.. These burial customs resemble to a great extent the situation in Mycenae. The swords were of new types, C- and D-type swords. They were shorter and better suited to warfare than the Neopalatial A-type swords. They are of an exceptionally high technical standard and were probably all made in the Knossian weapon workshop. The spearheads and knives were longer and the bronze vessels smaller than the Neopalatial ones. New object types appeared, *e.g.*, mirrors. Razors, ornaments and other personal objects, already common in the Early Minoan tombs but rare during the Neopalatial period, reappear in the archaeological record. Cult objects of bronze are rarely found.

The reasons for these changes are obvious: the shift from settlement finds to tomb finds, the military character of the "Mycenaeans", the depopulation of Crete and eastern Crete being almost deserted, the stagnate building activities and the new burial customs and ritual habits.

The manufacture of the large number of weapons attested on the Linear B Tablets of the Ra series (Macdonald 1997: 294), must have required huge amounts of copper and tin. The Knossian weapon workshop supposedly used only new, virgin metal, but no copper ingots from this period have been found. Based on finds from both the Neopalatial and the Postpalatial periods, one can suggest that the copper come from Cyprus. Lead isotope analyses for objects from the Unexplored Mansion at Knossos does not, however, support this hypothesis. They indicate that the source of the copper was Laurion (Stos-Gale and Gale 1990: 82). The objects analyzed were, however, probably old. The copper ingots hidden in a cellar in Hagia Triadha and under the LM IB destruction level at Zakros remained astonishly untouched. The new rulers apparently did not knew of their existence. But perhaps they found other ingots, *e.g.*, in Knossos?

Convincing evidence for the use of recycled scrap metal has been found in the LM II bronze workshop in the Unexplored Mansion at Knossos. It might indicate some shortage of metal but the ingots might also have been removed before the destruction. The workshop equipment, particularly the many crucibles, is evidence for a new bronze technology. The reasons for this shift is unknown; either was it a concious technological development or a different traditional craftmanship of the new rulers.

5.4 The Postpalatial period (LM IIIA2 - LM IIIC)

In the beginning of the Postpalatial period the "Mycenaeans" extended their territorial control and the deserted settlements were repopulated. Coastal centres such as Malia, Palaikastro, Hagia Triadha, Kommos and Chania fluorished and were important parts of the empire. After a recession, most of these sites were abandoned, but, *e.g.*, Chania, Phaistos and Knossos continued. After the final destruction of Knossos in LM IIIB2, the remaining "palatial" centre was Chania, with its commercial contacts to the western Mediterranean. In LM IIIC, defensible sites were founded in the mountains.

The published bronze objects from this long, dramatic period are too few, only 262, to enable any far-reaching conclusions about the bronzeworking. They indicate clearly, however, that almost all of Crete was repopulated. The majority of the bronze objects are found in graves all over the island, but the object assemblage in the graves is no longer so standardized, an indication of a weaker central control and that burials with objects were not solely for an elite. New types of weapons occur which were probably manufactured locally. Some swords of Knossian type indicate, however, that the weapon workshop in the palace was still functioning, but, of course, the swords could be heirlooms. The number of swords (42) exceed the number from the Mycenaean Knossos period. For many of them, however, the provenience and perhaps also the dating are uncertain. Six Peschiera type daggers from Italy belong to the extremely few bronze objects imported from outside the Aegean during the Late Bronze Age. The 28 vessels are all from tombs, 12 of them from the Mycenaean Burial Enclosure in Archanes. As during the previous period, personal objects as razors, ornaments and mirrors are normally found in the graves. But as a general rule, the grave gifts were not as luxurious as during the Mycenaean Knossos period.

The most important copper source was apparently Cyprus, but there are also indications that Laurion copper was still used. However, the objects might have been made of scrap metal. Interestingly, some objects from Chania are supposed to have been made of Sardinian copper. After the destruction of the palaces, bronzeworking continued, probably on a less sophisticated level on a smaller scale, in the coastal towns. Convincing evidence has been found in the settlement areas in Kommos. The metal-working equipment found differs markedly from the Neopalatial equipment from the southern area on the same site: the crucibles were hemispherical spouted low bowls of the same type as in the Unexplored Mansion and the moulds were clay investment moulds for lost-wax casting of, primarily, double-axes. Blitzer suggests that this production of double-axes was intended for external consumption (1995: 531). Finds of crucible, mould and tuyére fragments in a LM IIIB pit with metallurgical debris in Palaikastro support this shift in bronzeworking technology.

Chapter 6. Summary and conclusions

This study is the first attempt to present a comprehensive view of the entire bronze industry on Late Minoan Crete and its development from the Neopalatial to the Postpalatial period. For the first time, all published data related to Late Minoan bronzeworking have been collected from different publications and stored in databases. The 1857 bronze objects, 75 copper ingots or fragments and pieces of 157 metalworking equipment are characterized by coded parameters, which enable a wide range of analyses. For this study, however, I have chosen only to compare the data for the Neopalatial (MM III - LM I), the Mycenaean Knossos (LM II - LM IIIA1) and the Postpalatial (LM IIIA2 - LM IIIC) periods and identify differences between them in order to recognize possible developmental trends in the Late Minoan bronzeworking. The reasons for these changes and their relations to the social development on Crete are briefly discussed.

The roughly estimated amount of metal in the bronze objects in the database is one metric tonne, representing about 30 copper oxhide ingots or about the same number of ingots which have been found. This is an indication of the small proportion of bronzes which have survived. As this amount is by no means representative for the whole Late Minoan production of bronze objects, only conclusions on a very general level can be drawn from analyzing the data. Moreover, the data are to some extent non-compatible as they are stored "as published". For a full understanding of the Late Minoan "metallurgical landscape", archaeometallurgical and metallographic investigations, which so far are lacking, are crucial. But even the following general conclusions give some indications of the situation and can thus form a base for further investigations.

i) The main changes in Late Minoan bronzeworking occurred after the Neopalatial period, when the "Mycenaeans" had taken control of Knossos, concern the structure of the object assemblages, the object types, the find contexts, the bronze technology and the copper sources. The data from the Postpalatial period are insufficient for an adequate comparison.

ii) The object assemblages indicate that the priority for the use of the valuable metal apparently changed. During the Neopalatial period, the bulk of the metal was used for tools, e.g., heavy double-axes and saws for massive building programmes and shipbuilding. However, during the Mycenaean Knossos period, the manufacturing seems to have been concentrated on weapons of excellent craftmanship. Some object types almost disappeared, such as the slender triangular daggers with three rivets, the huge saws and cult objects. Instead, new object types such as long knives, short-swords, razors and mirrors appeared in the archaeological record. New weapons, better suited for warfare, were introduced: shorter C- and D-type swords and longer spearheads, either due to technological development or to a new ethnic group. These weapons were probably manufactured in the Knossian weapon workshop. During the Postpalatial period, the situation resembles the previous period, but some new object types occur: e.g., the Naue swords and the Peschiera daggers, the latter imported from Italy.

iii) The majority of the Neopalatial bronze objects were settlement finds from palaces and towns in eastern and central Crete. A considerable number of them come from bronze collections in LM IB destruction levels. During the Mycenaean Knossos period, almost all bronze objects are burial finds from the Knossos area. The objects from the Postpalatial period are also mainly from tombs, but they originate all over Crete, including western Crete.

iv) The evidence for bronze workshops in the palaces is rather weak, possibly due to the fact that it was hardly noticed in older excavations. In new excavations in coastal towns, convincing evidence for bronzeworking from all periods has fortunately been found. The most thoroughly investigated bronze workshop is, however, the LM II workshop in the Unexplored Mansion at Knossos.

v) The surprisingly few metallographic investigations and the chemical analyses published indicate that the bronzesmiths had excellent metallurgical skills for achieving the intended alloy composition and the right properties and physical form for the object. Only tin bronze and, for some object types, pure copper were used; arsenical bronze no longer occurred. Besides virgin copper and tin, re-melted scrap metal seems to have been commonly used in all periods.

vi) The Neopalatial casting technique, with large, deep, bowl-like clay crucibles on a high, pierced stem and stone moulds for mass production was replaced during the Mycenaean Knossos period by lower, hemispherical, bridge-spouted crucibles of a different fabric and clay investment moulds for lost-wax casting. The intended properties, i.e., increased thermal shock resistance, were achieved by different manufacturing techniques in these two clay types. The reason for this shift is not known.

vii) In all probability, the Minoans had to import all their metal, as the Cretan copper ore sources were quite insufficient. With their involvement in the international metal-trading network in the eastern Mediterranean, they had the opportunity to obtain copper from all available sources, of which some so far are unidentified. The major sources changed during different periods. The Neopalatial sources were mainly Laurion, an unknown source and to some extent Cyprus. From the Mycenaean Knossos period on, Cyprus seems to have been the dominant copper source and even Sardinian copper has been found from the Postpalatial period. Bulk copper was traded as oxhide ingots, and to some extent as bun ingots. There have been 30 intact and 39 fragments of oxhide ingots found on Crete, more than anywhere else in the Mediterranean, excluding the shipwrecks.

The Cretan Neopalatial ingots, dated to LM IA, are the oldest. Most of these were found hidden in a cellar at Hagia Triadha. In the bronze workshops, only fragments of ingots together with scrap metal have been found, an indication that the palaces probable controlled the importation of metals and distributed only ingot fragments to the workshops in the towns.

viii) The results of this study support the few published views of the Late Minoan bronze industry. The dramatic events on Crete during this period are clearly reflected in the bronzeworking, but based on available data, is it impossible to draw any far-reaching conclusions about the social aspects of bronze and its direct impact on the development of the society.

Bibliography

Alexiou, S.
1967 Υστερομινωικοί τάφοι λιμενός Κνωσού (Κατσάμπα). Βιβλιοθήκη της εν Αθηναίς Απχαιολογικής Εταιρείας, 56, Αθήνα.

Andreadaki-Vlasaki, M.
1997 La Nékropole du Minoen Récent de la Ville de la Canée. In *La Crète Mycénienne*, edited by J. Driessen and A. Farnoux. Bulletin de Correspondance Hellénique, Supplément 30, pp. 487 - 509. École francaise d'Athènes.

Avila, R. A. J.
1983 Bronzene Lanzen- und Pfeilspitzen der griechischen Spätbronzezeit. *Prähistorische Bronzefunde*, Abteilung V, Band 1. C. H. Beck'sche Verlagsbuchhandlung, Muenchen.

Baboula, E. and P. Northover
1999 Metals Technology versus Context in Late Minoan Burials. In Metals in Antiquity, edited by S. M. M. Young, A. M. Pollard, P. Budd and R. A. Ixer. *BAR International Series* 792: 146-152.

Banou, E. and G. Rethemiotakis
1997 Centre and Periphery: New Evidence for the Relations between Knossos and the Area of Viannos in the LM II - IIIA Periods. In *La Crète Mycénienne*, edited by J. Driessen and A. Farnoux. Bulletin de Correspondance Hellénique, Supplément 30, pp. 23-57. École francaise d'Athènes.

Bassiakos, Y.
2001 Early Copper Production on Kythnos: Materials and Analytical Reconstruction of Metallurgical Processes. *Sheffield Studies in Aegean Archaeology*. Sheffield Academic Press.

Bassiakos, Y. and P. P. Betancourt
2002 Early Copper Production in Southern Aegean: New Data. In *Early Metallurgy in Cyprus. The Last Twenty Years, 1982-2002*. Nicosia September 2002. Abstract.

Betancourt, P. P.
1998 Middle Minoan Objects in the Near East. In The Aegean and the Orient in the Second Millenium, edited by E. H. Cline and D. Harris-Cline. Proceedings of the 50th Anniversary Symposium Cincinnati 18-20 April 1997. Annales d'archéologie égéenne de l'Université de Liège et UT-PASP, *AEGEUM* 18:5-13.

2001 Excavations at Crysokamino, Crete. Reconstructing EM III Copper Smelting Processes. *9th International Congress of Cretan Studies*. Elounda 1-6 October 2001. Abstracts 14-15.

Betancourt, P., Y. Bassiakos and M. Catapotis
2003 Technological Study of Metallurgical Remains from the Prehistoric Smelting Site of Chrysokamino on Eastern Crete: Preliminary Results. *4th Symposium on Archaeometry of The Hellenic Society of Archaeometry*. Athens 28-31 May 2003. Book of Abstracts, 125.

Betancourt, P. P., J. M. Muhly, W. R. Farrand, C. Stearns, L. Onyshkevych, W.B. Hafford and D. Evely
1999 Research and Excavation at Chrysokamino, Crete, 1995 - 1998. *Hesperia* 68:343-370.

Betancourt, P. P., T. S. Wheeler, R. Maddin and J. D. Muhly
1978 Metallurgy at Gournia. *MASCA Journal* 1:7-8.

Blitzer, H.
1995 Minoan Implements and Industries. In *KOMMOS I. The Kommos Region and Houses of the Minoan Town. Part 1. The Kommos Region, Ecology and Minoan Industries*, edited by J. W. and M. Shaw, pp. 403-536. Princeton.

Boardman, J.
1961 *The Cretan Collection in Oxford. The Dictaean Cave and Iron Age Crete*. Oxford.

Bosanquet, R. C.
1901- Excavations at Palaikastro I. *The Annual of*
1902 *the British School at Athens* 8:286-316.

Boyd Hawes, H.
1908 *Gournia,, Vasiliki and other Prehistoric Sites on the Isthmus of Hierapetra, Crete*. Philadelphia

Branigan, K.
1968 Copper and Bronze Working in Early Bronze Age Crete. *Studies in Mediterranean Archaeology* 14. Lund.
1969 Early Aegean Hoards of Metalwork. *The Annual of the British School at Athens* 64:1-11.
1974 *Aegean Metalwork of the Early and Middle Bronze Age*. Oxford Monographs of Classical Archaeology, Oxford.
1983 Craft Specialization in Minoan Crete. In *Minoan Society*, edited by O. Krzyszkowska and L. Nixon. Proceedings of the Cambridge Colloquium 1981, pp. 23-32. Bristol Classical Press, Bristol.

Buchholz, H-G.
1959 Keftiubarren und Erzhandel im zweiten vorchristlichen Jahrtausend. *Prähistorische Zeitschrift* 37 (1/2):1-40.
1959a *Zur Herkunft der kretischen Doppelaxt: Geschichte und auswärtige Bezieungen eines minoischen Kultsymbols.* Muenchen.

Budd, P., A. M. Pollard, B. Scaife and R. G. Thomas
1995 Oxhide ingots recycling and the Mediterranean metals trade. *Journal of Mediterranean Archaeology* 8(1):1-32.

Catling, E. A. and H. W. Catling
1974 The Bronzes. In Sellopoulo Tombs 3 and 4. Two Late Minoan Graves near Knossos, by M. R. Popham. *The Annual of the British School at Athens* 69:225-254.

Catling, H. W.
1964 *Cypriot Bronzework in the Mycenaean World.* Claredon Press, Oxford.
1968 Late Minoan Vases and Bronzes in Oxford. *The Annual of the British School at Athens* 63:89-131.
1997 Minoan Metalworking at Palaikastro: some Questions. *The Annual of the British School at Athens* 92:52-58.

Catling, H. W. and E. Catling
1984 The Bronzes and Metalworking Equipment. In *The Minoan Unexplored Mansion at Knossos, Text,* by M. R. Popham et al. The British School of Archaeology at Athens, Supplementary Volume 17: 203-221. Thames and Hudson, Oxford.

Catling, H. W. and R. E. Jones
1976 Sellopoulo tomb 4: Some Analyses.*The Annual of the British School at Athens* 71:21-23.
1977 Analyses of copper and bronze artefacts from the Unexplored Mansion, Knossos. *Archaeometry* 19(1): 57-66.

Chapouthier, F. and P. Demargne
1942 Fouilles exécutées á Mallia. Troisiéme rapport: Exploration du palais (1927-1932). *Études Crétoises,* Tomé VI. École francaise d'Athénes. Athéne.

Charles, J. A.
1968 The First Sheffield Plate. *Antiquity* XLII: 278- 285.

Craddock, P. T.
1976 The Composition of the Copper Alloys used by the Greek, Etruscan and Roman Civilization. 1. The Greeks before the Archaic Period. *Journal of Archaeological Science* 3:93-113.

Dawkins, R. M.
1904- Excavations at Palaikastro. IV. *The Annual*
1905 of *the British School at Athens* 11:258-292.

Dawkins, R. M. and M. N. Tod
1902- Excavations at Palaikastro. II. *The Annual*
1903 *of the British School at Athens* 9:290-335.

Dercksen, J. G.
1996 The Old Assyrian Copper Trade in Anatolia. *Nederlands Historisch-Archaeologisch Instituut te Istanbul.* LXXV. Holland.

Deshayes, J.
1960 *Les Outils de Bronze, de l'Indus au Danube (IVe au IIe millénaire).* Paris.

Dimopoulou, N.
1997 Workshops and Craftsmen in the Harbour-Town of Knossos at Poros-Katsambas. In TEXNH. Craftsmen, Craftswomen and Crafts-manship in the Aegean Bronze Age, Vol. II, edited by R. Laffineur and P. P. Betancourt. Proceedings of the 6th International Aegean Conference, Philadelphia, Temple University, 18-21 April 1996. Annales d'archéologie égéenne de l'Université de Liège et UT-PASP, *AEGEUM* 16:433-438.

Driessen, J. and C. F. Macdonald
1984 Some Military Aspects of the Aegean in the Late Fifteenth and Early Fourteenth Century B.C. *The Annual of the British School at Athens* 79:49-74.
1997 The Troubled Island. Minoan Crete before and after the Santorini Eruption. Annales d'archéologie égéenne de l'Université de Liège et UT-PASP, *AEGEUM* 17.

Evans, A. J.
1906 *The Prehistoric Tombs of Knossos. I. The Cemetery of Zapher Papoura. II. The Royal Tomb of Isopata.* London.
1914 *The Tomb of the Double Axes and Associated Group and Pillar Rooms and Ritual Vessels of the "Little Palace" at Knossos.* London.

Evely, R. D. G.
1993 Minoan Crafts: Tools and Techniques. An Introduction. Vol I. *Studies in Mediterranean Archaeology* 92.1. Jonsered.
2000 Minoan Crafts : Tools and Techniques. An Introduction.Vol II. *Studies in Mediterranean Archaeology* 92.2. Jonsered.

Faure, P.
1966 Les Minerais de la Créte Antique. *Revue Archeologique* 45-78.
1968 Le Probleme du Cuivre dans la Créte Antique. Πεπραγμένα Β΄ Διεθνούς Κρητολογικού Συνεδρίου Β΄, pp. 174-193.

1980   Les Mines du Roi Minos. Πεπραγμένα Δ' Διεθνούς Κρητολογικού Συνεδρίου Α', pp. 150-168.

Forsdyke, E. J.
1926   The Mavro Spelio Cemetery at Knossos. *The*
-1927   *Annual of British School at Athens* 28:243-296.

Gale, N. H.
1989   Archaeometallurgical studies of Bronze Age copper oxhide ingots from the Mediterranean region. In Old World Archaeometallurgy, edited by A. Hauptmann, E. Pernicka and G.A. Wagner. *Der Anschnitt Beiheft* 7: 247-268. Deutsches Bergbau-Museum. Bochum.
1990   The Provenance of Metals for Early Bronze Age Crete - Local or Cycladic? Πεπραγμένα ΣΤ' Διεθνούς Κρητολογικού Συνεδρίου Α', pp. 299-316.
1991   Copper oxhide ingots, their origin and their place in the Bronze Age metals trade in the Mediterranean. In Bronze Age Trade in the Mediterranean, edited by N. H. Gale, *Studies in Mediterranean Archaeology* 90:197-239. Jonsered.
1998   The role of Kea in metal production and trade in the Late Bronze Age. In Kea - Kythnos: History and Archaeology, edited by L. G. Mendoni and A. Mazarakis Ainian. Proceedings of an International Symposium Kea - Kythnos, 22-25 June 1994. *ΜΕΛΕΤΗΜΑΤΑ* 27:738-752. Diffusion de Boccard / Paris, Athens.
1999   Lead Isotope Characterization of the Ore Deposits of Cyprus and Sardinia and its Application to the Discovery of the Sources of Copper for the Late Bronze Age Oxhide Ingots. In Metals in Antiquity, edited by S. M. M. Young, A. M. Pollard, P. Budd and R. A. Ixer. *BAR International Series* 792: 110-121.

Gale, N. H. and Z. A. Stos-Gale
1982   Bronze Age copper sources in the Mediterranean. A new approach. *Science* 216:11-19.
1986   Oxhide Copper Ingots in Crete and Cyprus and the Bronze Age Metals Trade. *The Annual of the British School at Athens* 81:81-100.
1986a   Recent Evidence for a Possible Bronze Age Metal Trade between Sardinia and the Aegean. In *Problems in Greek Prehistory,* edited by E. B. French and K. A. Wardle, pp. 349-384. Bristol.

1987   Oxhide Ingots from Sardinia, Crete and Cyprus and the Bronze Age Copper Trade. New Scientific Evidence. In Studies in Sardinian Archaeology III, Nuraic Sardinia and the Mycenaean World, edited by M. S. Balmuth. *BAR International Series* 387:135-177.

1992   Lead Isotope Studies in the Aegean. *Proceedings of the British Academy* 77:63-108.

Gale, N. H., Z. A. Stos-Gale, G. Maliots and N. Annetts
1997   Lead isotope data from the Isotrace Laboratory, Oxford. Archaeometry data base 4., Ores from Cyprus. *Archaeometry* 39:237-246.

Γεωργιάδου-Δικαιουλία, Ε., Μ. Σταματάκι and Μ. Δερμιτζάκης
1995   Συμβόλη στη γνώση της γεωλογίας των κυριότερων ορυκτών πρώτων υλών της νίσου Κρήτης. Πεπραγμένα του Ζ' Διεθνούς Κρητολογικού Συνεδρίου Α1', pp. 111 - 123.

Georgiou, H.
1979   The Late Minoan I Destruction of Crete. Metal Groups and Stratigraphic Considerations. *Monograph IX, Institute of Archaeology.* University of California, Los Angeles.

Gillis, C.
1993   Trade in the Bronze Age. In Trade and Production in premonetary Greece. Aspects of Trade, edited by C. Gillis, C. Risberg and B. Sjöberg. Proceedings of the Third International Workshop, Athens, *Studies in Mediterranean Archaeology and Literatur. Pocket Books* 134: 61-86. Jonsered.
1997   The Smith in the Late Bronze Age: State Employee, Independent Artisan or Both? In TEXNH. Craftsmen, Craftswomen and Craftsmanship in the Aegean Bronze Age, Vol. II, edited by R. Laffineur and P. P. Betancourt. Proceedings of the 6th International Aegean Conference, Philadelphia, Temple University, 18-21 April 1996. Annales d'archéologie égéenne de l'Université de Liège et UT-PASP, *AEGEUM* 16:505-513.

Hadjianastasiou, O. and S. MacGillivray
1988   An Early Bronze Age Copper Smelting Site on the Aegean Island of Kythnos, II. Archaeological Evidence. In *Aspects of Ancient Mining and Metallurgy,* edited by F. J. Jones. Acta of a British School at Athens Centenary Conference at Bangor, 1986, pp. 31-34.

Halbherr, F., E. Stefani and L. Banti
1980   Hagia Triada nel Periodo Tardo Palaziale. *Annuario della Scuola Archeologica di Atene e delle Missioni Italiane in Oriente* 55.

Hauptmann, A., R. Maddin and M. Prange
2002   Copper and Tin Ingots from the Shipwreck of Uluburun. In *Early Metallurgy in Cyprus. The Last Twenty Years, 1982-2002.* Nicosia September 2002. Abstract.

Hazzidakis, J.
1912 An Early Minoan Sacred Cave at Arkalokhori in Crete. *The Annual of the British School at Athens* 19:35-47.

Hemingway, S.
1996 Minoan Metalworking in the Postpalatial Period: a Deposit of Metallurgical Debris from Palaikastro. *The Annual of the British School at Athens* 91:214-243.
1999 Copper and Bronze Objects from Minoan Pseira. In Meletemata. Studies in Aegean Archaeology presented to Malcolm H. Wiener as he enters his 65th year. II, edited by P. P. Betancourt, V. Karageorghis, R. Laffineur and W-D.Niemeier. Annales d'archéologie égéenne de l'Université de Liège et UT-PASP, *AEGEUM* 20:357-360.

Hitchcock, L. A.
2001 Cult, Context and Copper: A Cypriot Perspective on the Unexplored Mansion at Knossos. *9th International Congress of Cretan Studies,* Elounda 1-6 October 2001.

Hood, M. S. F.
1956 Another Warrior-Grave at Ayios Ioannis near Knossos. *The Annual of the British School at Athens* 51:81-99.

Hood, M. S. F., G. Huxley and N. Sandars
1958- A Minoan Cemetery on Upper Gypsades.
1959 *The Annual of the British School at Athens* 53-54:194-253.

Hood, M. S. F. and P. de Jong
1952 Late Minoan Warrior-Graves from Ayios Ioannis and the New Hospital Site at Knossos. *The Annual of the British School at Athens* 47:243-277.

Hutchinson, R. W.
1956 A Late Minoan Tomb at Knossos. *The Annual of the British School at Athens* 51:68-73.
1956a A Tholos Tomb on the Kephala. *The Annual of the British School at Athens* 51:74-80.

Höckmann, O.
1980 Lanze und Speer in Spätminoischen und Mykenischen Griechenland. *Jahrbuch des Römisch-Germanischen Zentralmuseums. Mainz* 27:13-158.

Junghans, S., E. Sangmeister and M. Schröder
1968 *Studien zu den Anfängen der Metallurgie.* Berlin.

Kallitsaki, H.
1997 The Mycenaean Burial Enclosure in Phourni, Archanes. In *La Crète Mycénienne,* edited by J. Driessen and A. Farnoux. Bulletin de Correspondance Hellénique, Supplément 3 0, pp. 213-227. École francaise d'Athènes.

Kanta, A.
1980 The Late Minoan III Period in Crete. A Survey of Sites, Pottery and their Distribution. *Studies in Mediterranean Archaeology* 58. Göteborg.

Kassianidou, V.
1999 Bronze Age copper smelting technology in Cyprus - the evidence from Politico Phorades. In Metals in Antiquity, edited by S. M. M. Young, A.M. Pollard, P. Budd and R. A. Ixer. *BAR International Series* 792: 91-97.

Kilian-Dirlmeier, I.
1985 Noch einmal zu den 'Kriegergräbern'' von Knossos.*Jahrbuch des Römisch--Germanischen Zentralmuseums. Mainz* 32:196-213.
1993 Die Schwerter in Griechenland (ausserhalb der Peloponnes, Bulgarien und Albanien). *Prähistorische Bronzefunde* Abteilung IV, Band 12. Franz Steiner Verlag, Stuttgart.

Knapp, A. B.
1986 Copper Production and Divine Protection. Archaeology, Ideology and Social Complexity on Bronze Age Cyprus.*Studies in Mediterranean Archaeology and Literatur. Pocket Books* 42. Jonsered.
1990 Ethnicity, Entrepreneurship and Exchange. Mediterranean Inter-island Relations in the Late Bronze Age. *The Annual of the British School at Athens* 85:115-153.

Knapp, A. B. and J. F.Cherry
1991 Archaeological Science, Statistics and Cultural Solutions. Trade Patterns in the Bronze Age Eastern Mediterranean. *Archaeometry '90.* Int. Symposium on Archaeometry 2-6 April 1990. Heidelberg.
1994 Provenience Studies, and Bronze Age Cyprus. Production, Exchange and Politico-economic Change. *Monographs in World Archaeology* 21. Prehistory Press, Madison Wisconsin.

Knapp, A. B., M. Donnelly and V. Kassianidou
1998 Excavations at Politiko-Phorades 1997. *Report of the Department of Antiquities, Cyprus* 1998, pp. 247-268.

Knapp, A. B., J. D. Muhly and P. M. Muhly
1988 The Hoard is Human. Late Bronze Age Deposits in Cyprus and the Aegean. *Report of the Department of Antiquities, Cyprus* 1988, pp. 233 - 262.

Kontorli-Papadopoulou, L.
1984 Objets de l'age du bronze au Musee Bénaki. *Bulletin de Correspondance Hellénique* CVIII, pp. 13-25.

Luce, J. V.
1998 Late Bronze Age Trade, and the Homeric Tradition. *Report of the Department of Antiquities, Cyprus* 1998, pp. 55 - 66.

Macdonald, C.
1987 A Knossian Weapon Workshop in Late Minoan II and IIIA. In The Function of the Minoan Palaces, edited by R. Hägg and N. Marinatos. *Skrifter utgivna av Svenska Institutet i Athen* 4, XXXV, pp. 293-295.

MacGillivray, J. A.
1997 The Reoccupation of Eastern Crete in the Late Minoan II - IIIA 1/2 periods. In *La Crète Mycénienne,* edited by J. Driessen and A. Farnoux. Bulletin de Correspondance Hellénique, Supplément 30, pp. 275-279. École francaise d'Athènes.
2000 *Minotaur. Sir Arthur Evans and the archaeology of the Minoan Myth.* Jonathan Cape, London.

Mangou, H. and P. V. Ioannou
1997 On the Chemical Composition of Prehistoric Greek Copper-based Artefacts from the Aegean Region. *The Annual of the British School at Athens* 92:59-72.
1998 On the Chemical Composition of Prehistoric Greek Copper-based Artefacts from Crete. *The Annual of the British School at Athens* 93:91-102.
1999 On the Chemical Composition of Prehistoric Greek Copper-based Artefacts from Mainland Greece. *The Annual of the British School at Athens* 94:81-100.
2000 Studies of the Late Bronze Age Copper-based Ingots found in Greece. *The Annual of the British School at Athens* 95:207-217.

Μαραγκόυ, Λ. (editor)
1992 *Μινωιός και ελληνικός πολιτισμός. Απο την συλλογή Μητσοτάκη.* Ίδρυμα Γουλάνδρη, Μουσέιο Κυκλαδικές Τέχνης .Αθήνα.

Marinatos, S.
1935 Ausgrabungen und Funde auf Kreta. *Archäologischer Anzeiger,* pp. 247 - 254.

Matthäus, H.
1980 Die Bronzegefässe der kretisch - mykenischen Kultur. *Prähistorische Bronzefunde* Abteilung II, Band 1. C. H. Becksche Verlagsbuchhandlung, Muenchen.
1983 Minoische Kriegergräber. In *Minoan Society,* edited by O. Krzyszkowska and L. Nixon., Proceedings of the Cambridge Colloquium 1981, pp. 203-215. Bristol Classical Press.

Michailidou, A.
1995 Investigating metal technology in a settlement. The case of Akrotiri at Thera. *ΑΡΧΑΙΟΓΝΩΣΙΑ, Τόμος* 8, 1993 - 94.

1999 Η τεχνολογία του μετάλλου στην προϊστορική κοινωνία του Ακρωτηρίου Θήρας. 1 Διεθνές Συνεδρίο. Πράκτικα, pp. 645 - 651.

Muhly, J. D.
1973 Copper and Tin; The Distribution of Mineral Resources and the Nature of the Metals Trade in the Bronze Age. *Transactions of the Connectitut Academy of Arts and Science* 43:155-535.
1982 Lead isotope analysis and the Kingdom of Alashiya. *Report of the Department of Antiquities, Cyprus* 1982, pp. 210-218.

Muhly, J. D., R. Maddin and V. Karageorghis (editors)
1982 *Early Metallurgy in Cyprus 4000 - 500 B.C.* Acta of the International Arcaeological Symposium, Larnaca, Cyprus 1-6 June 1981. Pierides Foundation, Larnaca Cyprus.

Muhly, J. D., R. Maddin and T. Stech
1988 Cyprus, Crete and Sardinia. Copper Oxhide Ingots and the Bronze Age Metals Trade. *Report of the Department of Antiquities, Cyprus* 1988, pp. 281-298.

Nakou, G.
1995 The Cutting Edge: A New Look at Early Aegean Metallurgy. *Journal of Mediterranean Archaeology* 8(2):1-32.

Niemeier, W. D.
1983 The Character of the Knossian Palace Society in the Second Half of the Fifteenth Century B.C.: Mycenaean or Minoan? In *Minoan Society,* edited by O. Krzyszkowska and L. Nixon. Proceedings of the Cambridge Colloquium 1981, pp. 217-236. Bristol Classical Press, Bristol.

Northover, P. and D. Evely.
1995 Towards an appreciation of Minoan metallurgical techniques: information provided by copper alloy tools from Ashmolean Museum, Oxford. *The Annual of the British School at Athens* 90:83-105.

Nowicki, K.
2000 Defensible Sites in Crete c. 1200 - 800 B.C. (LM IIIB / IIIC through Early Geometric). Annales d'archéologie égéenne de l'Université de Liège et UT-PASP, *AEGEUM* 21.

Oberweiler, C., Y. Maniatis and J. Shaw
2003 Metallurgical Ceramics: Analysis on Crucibles and Moulds from Kommos, South Crete. *4th Symposium on Archaeometry of The Hellenic Society of Archaeometry.* Athens 28-31 May 2003. Book of Abstracts. 168.

Papadopoulos, Th. J.
1998 The Late Bronze Age Daggers of the Aegean Vol. 1. The Greek Mainland. *Prähistorische Bronzefunde,* Abteilung. VI, Band 11. Franz Steiner Verlag, Stuttgart.

Papadopoulou, E.
1997 Une Tombe à Tholos 'Intra Muros' le Cas du Cimetière MR d'Arménoi. In *La Crète Mycénienne,* edited by J. Driessen and A. Farnoux. Bulletin de Correspondance Hellénique, Supplément 30, pp. 319-340. École francaise d'Athènes.

Pelon, O.
1987 Minoan Palaces and Workshops: New Data from Malia. In The Function of the Minoan Palaces, edited by R. Hägg and N. Marinatos. *Skrifter utgivna av Svenska Institutet i Athen* 4, XXXV, pp. 269-272.

Pernicka, E.
1999 Trace Element Fingerprinting of Ancient Copper: A Guide to Technology or Provenance? In Metals in Antiquity, edited by S. M. M.Young, A. M. Pollard, P. Budd and R. A. Ixer. *BAR International Series* 792: 163-171.

Pernier, L. and L. Banti
1951 *Il Palazzo minoico di Festos, Vol. II. Il secondo palazzo.* Rome.

Platon, E. M.
1988 *The Workshops and Working Areas of Minoan Crete. The Evidence of the Palace and Town of Zakros for a Comparative Study.* Unpublished Dissertation at the University of Bristol.

Platon, N.
1974 *Ζάκρος. Το νέον μινωικόν ανάκτορον.* Αθήναι.
1979 'Lexportation du cuivre de l'ile de Chypre en Créte et les installations métallurgiques de la Créte minoenne. In *The Relations Between Cyprus and Crete, ca. 2000 - 500 B.C.* Acts of the International Archaeological Symposium, pp. 101-110.
1980 Μεταλλουργικό κάμινι στην Ζάκρο της Κρήτης. *Πεπραγμένα Δ' Διεθνούς Κρητολογικού Συνεδρίου* Α', 437 - 446.

Popham, M.R.
1974 Sellopoulo Tombs 3 and 4, two Late Minoan Graves near Knossos. *The Annual of the British School at Athens* 69:195-257.
1984 *The Minoan Unexplored Mansion at Knossos.* Two Volumes.The British School of Archaeology at Athens. Supplementary Volume 17. Thames and Hudson, Oxford.

Poursat, J-C.
1996 Artisans minoens: Les Maisons-Ateliers du quartier Mu. *Études Crétoises* 32. École francaise d'Athénes. Athéne.

Pulak, C.
2000 The Bronze and Tin Ingots from the Late Bronze Age Shipwreck at Uluburun. In Anatolian Metal I, edited by Unsal Yalcin. *Der Anschnitt Beiheft* 13: 137-157. Deutsches Bergbau-Museum. Bochum.

Pålsson Hallager, B.
1985 Crete and Italy in the Late Bronze Age III Period. *American Journal of Archaeology* 89:293-305.

Rehak, P. and J. G. Younger
1998 Review of Aegean Prehistory VII: Neopalatial, Final Palatial and Postpalatial Crete. *American Journal of Archaeology* 102:91-173.

Rehren, Th. and J. P. Northover
1991 Selenium and Tellurium in ancient copper ingots. *Archaeometry* '90. International Symposium on Archaeometry Heidelberg 2-6 April 1990, pp. 221-228.

Renfrew, C.
1967 Cycladic metallurgy and the Aegean Early Bronze Age. *American Journal of Archaeology* 71:1-20.
1972 *The Emergence of Civilisation. The Cyclades and the Aegean in the Third Millenium B.C.* Methuen & Co. Ltd, London.

Sackett, L. H. and M. R. Popham
1963 Excavations at Palaikastro. VI. *The Annual of the British School at Athens* 60:247-315.

Sakellarakis, Y. and E. Sakellarakis
1997 *Archanes, Minoan Crete in a New Light.* Volumes I and II. Ammos Publications, Eleni Nakou Foundation.

Sandars, N. K.
1955 The Antiquity of the One-edged Bronze Knife in the Aegean. *Proceedings of the Prehistoric Society* 21:174-197.
1961 The First Aegean Swords and their Ancestry. *American Journal of Archaeology* 65:17-29.
1963 Later Aegean Bronze Swords. *American Journal of Archaeology* 67:117-153.

Sapouna-Sakellarakis, E.
1995 Die bronzenden Menschenfiguren auf Kreta und in der Ägais. *Prähistorische Bronzefunde* Abteilung I, Band 5. Franz Steiner Verlag, Stuttgart.

Shaw, J. W.
1973 Minoan Architecture: Materials and Techniques. *Annuario della Scuola Archeologica di Atene e delle Missioni Italiane in Oriente.* Vol. XLIX.

Shaw, J. W. and M. C. Shaw (editors)
1995 *Kommos I. The Kommos Region and Houses of the Minoan Town. Part 1. The Kommos Region, Ecology an*d *Minoan Industries.* Princeton.

Shaw, J. W. and M. C. Shaw
1997 "Mycenaean" Kommos. In *La Crète Mycénienne,* edited by J. Driessen and A. Farnoux. Bulletin de Correspondance Hellénique, Supplément 30, pp. 423-434. École francaise d'Athènes.

Skarpelis, N. and Y. Bassiakos
2000 The Aegean Epithermal-Type Mineralizations. A Source of Precious and Base Metals in Antiquity. In *5th Mining History Congress.* Milos 12-15 September 2000. Book of Abstracts, p. 105.

Soles, J. S.
1996 A Community of Craft Specialists at Mochlos. In TEXNH. Craftsmen, Craftswomen and Craftsmanship in the Aegean Bronze Age, Vol.II, edited by R. Laffineur and P. P.Betancourt. Proceedings of the 6th International Aegean Conference, Philadelphia, Temple University, 18-21 April 1996. Annales d'archéologie égéenne de l'Université de Liège et UT-PASP, *AEGEUM* 16:425-432.

Soles, J. S. and C. Davaras
1992 Excavations at Mochlos, 1989. *Hesperia* 61(4):413-445.
1994 Excavations at Mochlos, 1990-1991. *Hesperia* 63(4):391-436.
1996 Excavations at Mochlos, 1992-1993. *Hesperia* 65(2):175-230.
2001 A Cypriot Vase from LM IB Mochlos. *9th International Congress of Cretan Studies.* Elounda 1-6 October 2001. Abstracts, p.77.

Stos-Gale, Z. A.
1986 Lead Isotope Evidence for Trade in Copper from Cyprus during the Late Bronze Age. In *Problems in Greek Prehistory,* edited by E. B. French and K. A. Wardle, pp. 265-282. Bristol.
1989 Cycladic Copper Metallurgy. In Old World Archaeometallurgy, edited by A. Hauptmann, E. Pernicka and G. A. Wagner. *Der Anschnitt Beiheft* 7: 279-291.Deutsches Bergbau-Museum. Bochum.

1993 The Origin of Metal used for Making Weapons in Early and Middle Minoan Crete. In *Trade and Exchange in Prehistoric Europe,* edited by C. Scarre and F. Healy. Proceedings of a Conference held at the University of Bristol, April 1992, pp. 115 - 129. Oxbow Books in association with The Prehistoric Society and the Société Préhistorique Francaise.
1998 The Role of Kythnos and other Cycladic Islands in the Origin of Early Minoan Metallurgy. In Kea - Kythnos: History and Archaeology, edited by L. G. Mendoni and A. Mazarakis Ainian. Proceedings of an International Symposium Kea - Kythnos, 22-25 June 1994. *MEΛETHMATA* 27:717-735. Diffusion de Boccard/Paris, Athens.

Stos-Gale, Z. A. and N. H. Gale
1990 The Role of Thera in the Bronze Age Trade in Metals. In *Thera and the Aegean World III,* Volume One, Archaeology, edted by D A. Hardy with C. G. Doumas, J. A. Sakellarakis and P. M. Warren. Proceedings of theThird International Congress, Santorini, Greece 3-9 September 1989, pp. 72-92. The Thera Foundation, London.
1994 Metal. In Provenience Studies, and Bronze Age Cyprus. Production, Exchange and Politico-economic Change, by A. B. Knapp and J. F. Cherry. *Monographs in World Archaeology* 21:92-121. Prehistory Press, Madison Wisconsin.

Stos-Gale, Z. A., N. H. Gale and N. Annetts
1996 Lead isotope data from the Isotrace Laboratory, Oxford. Archaeometry data base 3., Ores from the Aegean. Part 1. *Archaeometry* 38:381-390.

Stos-Gale, Z. A., N. H. Gale and D. Evely
2000 An Interpretation of the Metals Finds, using Lead Isotope and Chemical Analytical Procedures. In The Greek-Swedish Excavations at the Agia Aikaterini Square. Kastelli, Khania, 1970 -1987. Vol. II, The Late Minoan IIIC Settlement, edited by E. Hallager and B. P. Hallager. *Skrifter utgivna av Svenska Institutet i Athen,*4, XLVII, II:206-214.

Stos-Gale, Z. A., N. H. Gale, J. Houghton and R. Speakman
1995 Lead isotope data from the Isotrace Laboratory, Oxford. Archaeometry data base 1., Ores from Western Mediterranean. *Archaeometry* 37:407-415.

Stos-Gale, Z. A., N. H. Gale and A. Papastamaki
1988 An Early Bronze Age Copper Smelting Site on the Aegean Island of Kythnos, I. Scientific Evidence. In *Aspects of Ancient Mining and Metallurgy,* edited by E. J. Jones. Acta of a British School at Athens Centenary Conference at Bangor, 1986, pp. 23-30.

Stos-Gale, Z. A., N. H. Gale and U. Zwicker
1986 The Copper Trade in the South-East Mediterranean Region. Preliminary Scientific Evidence. *Report of the Department of Antiquities, Cyprus* 1986, pp. 122-144.

Stos-Gale, Z. A. and C. F. Macdonald
1991 Sources of metals and trade in the Bronze Age Aegean. In Bronze Age Trade in the Mediterranean, edited by N. H. Gale, *Studies in Mediterranean Archaeology* 90:248-280.

Stos-Gale, Z.A., G. Maliots, N. H. Gale and N. Annetts
1997 Lead isotope characteristics of the Cyprus copper ore deposits applied to provenience studies of copper oxhide ingots. *Archaeometry* 39:83-123.

Tselios, T.
2003 A New Program on Minoan Metallurgical Techniques during the Early and Middle Bronze Age. *4th Symposium on Archaeometry of The Hellenic Society of Archaeometry.* Athens 28-31 May 2003. Book of Abstracts, 141.

Tylecote, R. F.
1981 From Pot Bellows to Tuyéres. *Levant,* XIII pp.107-118.

Varoufakis, G. J.
1995 Chemische und metallurgische Untersuchungen von 45 minoischen Statuetten. In Die Bronzenden Menschenfiguren auf Kreta und in der Ägais, by E. Sapouna-Sakellarakis. *Prähistorische Bronzefunde* Abteilung I, Band 5, pp.154-167. Franz Steiner Verlag, Stuttgart.

Watrous, L. V.
1994 Review of Aegean Prehistory III: Crete from Earliest Prehistory through the Protopalatial Period. *American Journal of Archaeology* 98:695-753.

Watrous, L.V. and H. Blitzer
1997 Central Crete in Late Minoan II - IIIB1. The Archaeological Background of the Knossos Tablets. In *La Crète Mycénienne,* edited by J. Driessen and A. Farnoux. Bulletin de Correspondance Hellénique, Supplément 3 0 , pp. 511-516. École francaise d'Athènes.

Wiener, M. H.
1990 The Isles of Crete? The Minoan Thalassocracy Revisited. In *Thera and the Aegean World* III, Volume One, Archaeology, edted by D A. Hardy with C. G. Doumas, J. A. Sakellarakis and P. M. Warren. Proceedings of theThird International Congress, Santorini, Greece 3-9 September 1989, pp. 128-161. The Thera Foundation, London.

Woodhead, A. P., N. H. Gale and Z. A. Stos-Gale
1999 An Investigation into the Fractionation of Copper Isotopes and its possible Application to Archaeometallurgy. In Metals in Antiquity, edited by S. M. M. Young, A. M. Pollard, P. Budd and R. A. Ixer. *BAR International Series* 792: 134-139.

Appendix I. Bronze workshops

Appendix I.1

Malia, Bronze workshops

Location	Palace on north coast. Two metalworking areas: the earlier in Quartier Mu NW of the palace and the later in the NW area of the palace.
Dating	MM II and MM III
	Quartier Mu, MM II
Material remains	1 small intact clay crucible.
	1 open stone mould for 4 tools.
	5 two-piece stone moulds for, *e.g.*, double-axes.
	4 tuyéres, fragments.
	Palace, NW area, MM III
	6 open stone moulds for tools / circular objects.
	6 two-piece stone moulds for double-axes.
	Burnt spot.
	Slag.

During the Protopalatial period, Malia has been considered an important metallurgical centre; numerous bronze objects have been found in the palace and its surroundings. The new evidence for metalworking in Quartier Mu strongly supports this hypothesis.

The bronze workshop in Quartier Mu is a well-defined house among a cluster of similar artisan establishments for, *e.g.*, ceramics, seals and stones. The two-storeyed building comprised both a habitation area and a workshop. The stone moulds and bronze tools found in the house had fallen from the upper floor. Nearby the workshop were found the tuyére fragments, and the only crucible. Analyses of the more than 30 bronze objects from Quartier. Mu, of which many might have been manufactured in the workshop, show that arsenical bronze was already mainly replaced by tin bronze; the weapons contained 8-10 % tin.

The deposit of stone moulds with clear traces of fire and repairs, found by Chapouthier in 1928 in the NW part of the palace, could not be tied to any structural remains. In the 1980s, however, a highly burnt spot mixed with ash and slag, which might be interpreted as the remains of smelting on a small-scale of rich copper ore in crucibles was discovered nearby. The workshop was located outside the MM III palace. In fact, the whole quarter could have been dedicated to craft activities.

References

Poursat   1996   Artisáns minoéns: Les maisóns-ateliérs du Quartièr Mu. *Études Crétoises* 32.

Pelon   1987   Minoan Palaces and Workshops: New Data from Malia. In *The Function of the Minoan Palaces*, edited by Hägg and Marinatos, pp. 269-272.

Evely   2000   Minoan Crafts. Tools and Techniques. Vol II. *SIMA* 92.2, 356, 358.

Appendix I.2

Gournia, Bronze workshops

Location Town by the Bay of Mirabello in eastern Crete. Three locations identified for metalworking: units Ea and Fh and house Cg.

Dating LM IB

Material remains 4 fragments of copper ingots.

Metal sheets, strips and scrap.

5 open stone moulds for chisels, nails and an axe?

A stone crucible ?

A hammer and other tools probably used in metal-working.

Slag?

Gournia was excavated in 1901-1904 by Boyd Hawes. The residents lived and practised a wide varity of crafts, of which one of the most important was metalworking. It is likely that Gournia was the source for many of the bronze tools and household utensils used by surrounding villages and farmsteads. Many bronze objects were found in the town and there is evidence that at least some of them were manufactured on the site.

In two units, Ea and Fh, more or less opposite each other across a street in the northern part of the town, a total of 5 open stone moulds where found. One, with matrices for 13 tools, had been carefully repaired with a copper wire. However, there is no complementary evidence for actual metalworking that has survived, unless the slight indication of slag and bronze in Fh. In House Cg, which seems to have belonged to a worker and dealer in bronze, a quantity of bronze scrap was found. No tools, no debris, nothing exists with which to pinpoint actual production here. A stone object of peculiar type, was identified by the excavator as a crucible, but now it is supposed to have had some other function.

Based on chemical and metallographical analyses of the ingot fragments and some small bronze objects, Betancourt *et al.* could not clearly conclude that these were associated. The objects contained small amounts of tin, which indicates that scrap was used in their manufacturing.

References
Boyd Hawes   1908   *Gournia.*. Philadelphia
Betancourt *et al.*   1978   Metallurgy at Gournia. *MASCA J.* 1, 7-8.
Evely   2000   Minoan Crafts: Tools and Techniques. Vol II *SIMA* 92.2, 335, 338.

Appendix I.3

Mochlos, Bronze workshop

Location Coast opposite Mochlos island in eastern Crete. Building A.

Dating LM IB - LM II

Material remains Bronze foundry hoard: 2 broken copper-alloy bowls containing fragments of copper ingots.

Run-off and waste from the casting process, numerous pieces.

2 clay investment moulds, probably for the handles of a large bowl.

Stone and pumice tools.

Evidence for freshly cast objects: impression of dagger in clay and small bronze objects such as a pin, knives, nails, fish hooks.

No crucibles, no bellows and no tongs were found.

The Greek-American excavations at Mochlos uncovered in 1990-1991 a workshop area on the coast opposite the island of Mochlos, which comprised workshops for bronze objects, stone vases, pottery and textiles. Each workshop area also accommodated cooking and eating facilities and they appear to have been used as residences as well as workshops. Each workshop may also have contained a small shrine. The area is unusual for several reasons: its location, its date and its organization.

The bronze workshop was located in building A. Material remains were found mainly in two rooms, but run-off and waste from the casting process existed in virtually every room. The ingot fragments are of copper from Cyprus. The hearths found look as if they were used for cooking. The actual casting process may have taken place outside, probably in the destroyed northern room which may have been open to the sea.

The artisans at Mochlos should probably be described as independent, full-time specialists. Standing apart from the main settlement at Mochlos, far from any palace, they were independent of any elite control. They were meeting a general economic demand for the goods they produced which was large enough to support them. Such a market could not be provided by the Mochlos settlement alone where there were additional artisans producing for local consumption. Their market would have been the Gournia polity around the Bay of Mirabello.

References.

Soles/Davaras 1994 Excavations at Mochlos 1990 - 91. *Hesperia* 63, 4: 391-436.

Soles 1996 A Community of Craft Specialists at Mochlos. TEXNE. *AEGEAN* 16, 425-432.

Soles/Davaras 2001 A Cypriote Vase from LM IB Mochlos. *9th Int. Congress of Cretan Studies*, Elounda. Abstracts, 77.

Appendix I.4

Zakros, Bronze workshop

Location   Coastal palace and town in eastern Crete. Industrial area in the south wing of the palace and near the harbour road.

Dating   LM IB

Material remains   6 intact copper oxhide ingots in the west wing.

1 clay crucible.

1 two-piece stone mould for a double-axe?

Fragments of sheet from bronze vessels (scrap?).

Problematic "fingered kiln" - metallurgical furnace or pottery kiln ?

The presence of a great number of bronze objects, and the discovery of bronze raw material in the form of intact copper oxhide ingots in the west wing, indicate that one or more bronze workshops functioned in the palace. Though it seems certain that the bronze objects were made in Zakros, there is not enough evidence directly related to bronzeworking. The south wing of the palace is the focus for several craft activities; bronzeworking is suggested to have taken place on the upper storey. Vessels, whole and parts thereof, turned up in room *χlνa*. The oxhide ingots fallen near rooms *xxiv-xxvi*, are an indisputable part of bronzeworking, but do not necessarily indicate that actual manufacture took part in the west wing. The presence of whole ivory tusks and raw stone nodules nearby rather indicate storage. Direct evidence for metalworking has only been found outside the palace where in the House of Niches grinding stones, charcoal, a clay crucible and a stone mould were found.

The "fingered kiln" near the harbour road outside the palace, was interpreted by N. Platon as a metallurgical kiln based on the "slag" found . The kiln is now restored. However, there is no scientifically supported association between the kiln and any metallurgical process, and presently there is a consensus among specialists that the Zakros kiln, as well as kilns of similar type in Phaistos and Hagia Triadha, are pottery kilns. The most telling argument lies in their large dimensions.

References

Platon, N.   1974   *Ζαχρος. Το νέον μινωικόν ανάκτορον.*
Platon, N.   1980   Μεταλλουργικό κάμινι στην Ζάχρος της Κρήτης. *Πεπραγμένα Δ΄ Διεθνούς Κρητολογικού Συνεδρίου Α΄*, 437-446.
Platon, E.M. 1988   *The Workshops and Working Areas of Minoan Crete.* Unpublished dissertation at the University of Bristol.
Evely   2000   Minoan Crafts. Tools and Techniques. Vol II. *SIMA* 92.2, 341.

Appendix I.5

Poros-Katsambas, Bronze workshops

Rescue excavations by Dimopoulou in the 1990s have clarified that Poros - Katsambas was a flourishing manufacturing and artistic centre from the Prepalatial to the Postpalatial period. Among the crafts, obsidian working, seal and jewelry making and metalsmithing were of significant importance. Evidence for metalworking has been found not only in pits and wells, but also in stratified contexts.

Evidence for copper / bronze work is traced first in a MM IIB context. A fragmentary but sizeable crucible and byproducts of copper melting were found at the Skatzourakis plot. During the Neopalatial period, bronzeworking extended all over the zone. In the Charonitakis plot near Kairatos, several crucible fragments, slag and melting droplets related to a tuyére of the straight cylindrical type set into an open channel were found.. In the Psychogioudakis plot, besides the evidence for metal melting common in the Neopalatial layers all over the area, remains of a crucible furnace were excavated: a small cavity cut into the rock and lined with clay. In the Sanoudakis plot, the debris of a two-storeyed house contained crucible fragments, pieces of copper ingots, other raw and waste material and two tuyéres. Metalworking continued on the same plot during the Postpalatial period from which a constructed rectangular furnace has been excavated. During the Neo- and Postpalatial periods the activities were extended to the Trypeti Hill, as attested by a significant find, an intact copper oxhide ingot.

The bronze smiths, with direct access to imported raw material and to the export trading system, were probably partially autonomous from the palatial bureaucratic centralization.

Location	Harbour town of Knossos. Evidence for metal-working in the residential areas in the central and southern outer zone of the settlement
Dating	MM II, MM III - LM I and LM IIIA2 - LM IIIB
Material remains	MM II
	1 sizeable crucible. Byproducts of copper melting.
	MM III - LM I (4 plots)
	Intact copper oxhide ingot. Pieces of copper ingots. Lead ingot. Numerous crucible fragments. 3 tuyéres. Bivalved clay mould and pot bellow (fragments). Slag, scrap, waste. Crucible furnace.
	LM IIIA2 - LM IIIB
	Constructed furnace filled with ash.

References

Dimopoulou 1997 Workshops and Craftsmen in the Harbour-Town of Knossos at Poros-Katsambas. TEXNE. *AEGEUM* 16, 433-438.

No detailed report has so far been published.

Appendix I.6

Knossos, The Unexplored Mansion, Bronze workshop

Location Building in the Knossos area, adapted rather than being custom-built as a workshop. Five rooms contain evidence of metalworking .

Dating LM II

Material remains Remains of a small furnace, usable as a smith´s melting furnace.

More than 300 objects of which over 200 copper and bronze objects are from LM II, comprising

- foundry equipment (*e.g.,* chisels, awls, punches, hammers and mould wires),
- scrap (*e.g.,* pieces of long saws and vessels),
- waste (*e.g.,* jets, flashers, risers).

1 small clay mould.

23 hemispherical clay crucibles.

No virgin raw materals or fuel were found.

The Unexplored Mansion at Knossos was excavated by Popham in 1972-1973. H. and E. Catling studied the metals and foundry equipment. The building, constructed in LM IA, was never completed. It was occupied in LM II, but not in the manner its planners envisaged; for bronzeworking. During the same period it was damaged by fire and the upper floor collapsed. The southern sector was left in ruins, but the northern sector was cleared out and reoccupied. Only this sector has been excavated.

Even incomplete, the Unexplored Mansion represents a bronze workshop on a much grander scale than the other workshops found so far on Crete. It is valuable evidence of the bronze industry shortly after the ”Mycenaeans” arrived at Knossos. Due to the distribution pattern of the objects, no local for the actual work can be discovered, but it is suggested that the work was conducted on the upper floor or possibly outside in a yard. The workshop is evidence of the importance of recycling metal, at least during LM II, perhaps due to a shortage of virgin copper and tin.

Analyses of the tin contents of 130 objects, including finished objects, scrap and waste, indicate some degree of compositional homogeneity for certain object categories as well as the controlled and economical use of tin.

References

Catling/Jones 1977 Analyses of copper and bronze artefacts from the Unexplored Mansion, Knossos. *Archaeometry,* 19: 1, 57-66.

Popham *et al.* 1984 The Minoan Unexplored Mansion at Knossos. *BSA Supplement,* Volume 17.

Catling *et al.* 1984 The Bronzes and Metalworking Equipment. In Popham 1984, 203-222.

Evely 2000 Minoan Crafts: Tools and Techniques. Vol II *SIMA* 92.2, 338.

Appendix 1.7

Kommos, Bronze workshops

Location	Harbour town for Mesara in southern Crete. There were metalworking activities during two distinct periods in different parts of the town.
Dating	MM III - LM I and LM II - LM III
Material remains	MM III - LM I
	11 pedestaled crucibles, fragments.
	1 copper ingot fragment.
	Metal bars, wire, strips intended for manufacturing.
	Slag, debris, pumice tools.
	Clay larnax for quenching.
	LM II - LM III
	Furnace bed or clay hearth.
	11 bowl-shaped crucibles, fragments.
	12 clay investment moulds for double-axes, fragments.
	3 clay pot bellows, fragments.
	5 copper ingot fragments, scrap.
	Metal bars, wire, strips intended for manufacturing.
	Slag, debris, pumice tools.

Kommos has been under excavation by an international team led by J. W. Shaw from the University of Toronto since 1976. The town on the shore consists of three different parts: the Hilltop and the Central Hillside which are settlement areas and a Civic Centre. Substantial evidence for metalworking activities has been found from two periods, which are all published in detail by Blizer.

In the MM III - LM I period, the metalworking activities were concentrated in the public buildings of the southern harbour complex. A full-scale metalmelting operation was in place in Building T, into the floor of which has been sunk a larnax for quenching. The floor was strewn with charcoal and fragments of crude chaff-tempered clay, pedestaled crucibles (dia. 25-30 cm). These crucibles tie into a technological pattern on the island from as early as EM. The crucibles contain metal droplets on their interiors, but there is no clear sign of the raw material, either copper ingots or scrap. No moulds were found in association with the crucibles.

Beginning at the end of LM I and definitely by LM II, metalworking appears to have ceased in the southern complex and began to be carried out within the surrounding town as small-scale industrial operations. The artifactual evidence consists primarily of bowl-shaped, bridge-spouted clay crucibles, which occur throughout Crete in this period, clay investment moulds for double-axes and massive clay pot bellows. New types of hearths were introduced consisting of upright stone blocks lined with clay. The production of double-axes might have been intended for external consumption.

Reference
Blizer   1995 Minoan Implements and Industries. In *Kommos* I, Part 1, edited by J. Shaw and M. Shaw, pp. 500-535.

Appendix I.8

Palaikastro, Bronze workshop

Location Large coastal town in eastern Crete. Metallurgical deposit consisting of debris from a small-scale melting and casting operation.

Dating LM IIIB

Material remains 3 hemispherical clay crucibles, fragments.

Numerous fragments of clay investment moulds, belonging to at least 12 moulds, mainly for tools.

8 tuyéres, fragments.

During excavations in 1991 in Palaikastro, a pit with metallurgical debris was found. As metallurgical evidence from the Postpalatial period is extremely rare on Crete, the debris was investigated in detail by Hemingway from the British School.

Three types of metallurgical evidence were identified in the debris: tuyéres, clay crucibles and clay investment moulds. No ingots, scrap or waste were found nor was slag or fuel. The three crucibles are of the same type as found in the Unexplored Mansion, but larger, with maximum diameters of 21 cm and maximum heights of 22 cm, or the largest found on Crete. In totals 84 mould fragments were found belonging to at least 12 clay investment moulds for lost-wax production, mainly of tools. Several of the moulds, however, Hemingway identified as intended for the casting of parts for bronze stands, so far not found on Crete but common on Cyprus. The moulds are of the same type as found in the LM III context at Kommos. The shape and size of the tuyére fragments are all close to Tylecote's type "diverted tuyére". These are the only tuyéres found in Crete in a LM context.

Hemingway saw the clay moulds as indicating a general shift in casting technique in the Postpalatial period; from stone moulds to clay moulds. Catling argues against this hypothesis, stating that the evidence so far is insufficient and that stone moulds were valuable goods which never would be discarded in a pit.

References

Hemingway 1996 Minoan Metalworking in the Postpalatial Period: A Deposit of Metallurgical Debris from Palaokastro. *BSA* 91, 214-243.

Catling 1997 Minoan Metalworking at Palaikastro: Some questions. *BSA* 92, 52-58.

Tylecote 1981 From Pot Bellows to Tuyéres. *Levant* XIII, 107-118.

Appendix II. Parameters and abbreviations

Appendix II.1

Parameters in databases (App. III, V)

			Nail, rivet	14
1.	**Number**	**No**	Tweezers	15
2.	**Site**	**Site**	Stake	16
3.	**Context**	**Co**	"Cone"	17
4.	**Geografical area**	**G**	Razor	18
	Unspecified	0	Cutter	19
	Eastern Crete	1	Other tool	20
	North Central Crete	2	Vessel	21
	South Central Crete	3	Scale pan	22
	Western Crete	4	Fish hook	29
5.	**Geografical location**	**L**	Sword	31
	Unspecified	0	Dagger	32
	Coastal	1	Spearhead	33
	Inland	2	Arrowhead	34
6.	**Site type**	**S**	Helmet	36
	Unspecified	0	Cult double-axe	45
	Palace	1	Figurine, male	46
	Palace surrounding	2	Figurine, female	47
	Town	3	Figurine, other	49
	Villa	4	Mirror	51
	Village	5	Ring	52
	Tomb	6	Bracelet	53
	Cave	7	Bead	54
	Sanctuary	8	Pendant	55
	Other	9	Earring	56
7.	**Period**	**Pe**	Other object	60
	Unspecified LM	0	Scrap	61
	Neopalatial	1	Waste	62
	Mycenaean Knossos	2	Mould wire	63

	Postpalatial	3	10.	**Object category**	**C**
8.	**Date**	**Da**		Unspecified	0
	Unspecified	0		Utilarian	1
	MM IIIA	1		Prestige	2
	MM IIIB	2		Cult	3
	LM IA	3	11.	**Typology, vessels**	**TM**
	LM IB	4		Unspecified	0
	LM II	5		Cauldron	1
	LM IIIA1	6		Tripod cauldron	2
	LM IIIA2	7		Broad pan	3
	LM IIIB	8		Pan	5
	LM IIIC	9		Kratere	6
9.	**Object type**	**Ot**		Amphora	7
	Unspecified	0		Hydria	8
	Chisel or oxhide ingot	1		Ewer	9
	Knife or bun ingot	2		Spouted jug	10
	Double-axe	3		Basin	11
	Single-axe	4		Cup	12
	Double-adze	5		Vapheio cup	13
	Axe-adze	6		Kylike	16
	Flataxe-adze	7		Lekanai	17
	Pick-adze	8		Bowl	18
	Saw	9		Ladle	19
	Drill	10		Lamp	20
	Hammer	11		Ash-tray	21
	Awl, point, punch	12		Sieve	22

Parameters in databases. (cont.)

12.	**Typology, swords**	**TK**
	Unspecified	0
	Sword, no rivetholes	1
	Type A	2
	Type B	4
	Type Ci	5
	Type Cii	6
	Type Gi	7
	Type Gii	8
	Type H	9
	Type Di	11
	Type Dii	12
	Type Fi	13
	Type Fii	14
	Unspecified fragments	15
	Type Naue II	16
13.	**Typology, daggers**	**TP**
	Unspecified	0
	Tangless type	1
	Dirk, shortsword	2
	Peschiera type	3
14.	**Condition**	**Cn**
	Unspecified	0
	Intact	1
	Nearly intact	2
	Fragment	3
15.	**Length, cm**	**L**
16.	**Width, cm**	**Wi**
17.	**Height, cm**	**He**
18.	**Rim diam., cm**	**Rd**
19.	**Thickness, cm**	**Th**
20.	**Weight, g**	**We**
21.	**Copper, %**	**Cu**
22.	**Tin, %**	**Sn**
23.	**Arsenic, %**	**As**
24.	**Lead, %**	**Pb**
25.	**Antimony, %**	**Sb**
26.	**Iron, %**	**Fe**
27.	**Nickel, %**	**Ni**
28.	**Cobalt, %**	**Co**
29.	**Zinc, %**	**Zn**
30.	**Bismuth, %**	**Bi**
31.	**Total metals, %**	**Tot**
32.	**Analysis method**	**Met**
32.	**Analysis date**	**AnDa**
33.	**Provenience for copper**	**Pr**
	No analysis	0
	Kythnos	1
	Laurion	2
	Cyprus	3
	Sardinia	4
	Unknown source	5
34.	**Museum or signum**	**Mu**
35,	**References**	**Ref**
36.	**Number in object database**	**DBno**

Grouping of object type codes for coarse classification

Tools	=	codes 1 - 17, 19 - 20
Vessels	=	codes 21 - 22
Weapons	=	codes 31 - 34, 36
Cult objects	=	codes 45 - 49
Personal objects	=	codes 18, 51 - 56
Other objects	=	codes 29, 60

Appendix II.2

Abbreviations and coding of sites. Total 137

Site	Abbr.	G	L	S	Site	Abbr.	G	L	S
Adromyloi	**Adr**	1	2	6	Knossos	**Kno**	2	2	1, 2
Ag-cup / tomb	**AgC**	2	2	6	Kommos	**Kom**	3	1	3
Akropoli grave	**Akr**	2	2	6	Kophinas	**Kop**	1	2	8
Amygdalois L.	**Amd**	1	2	0	Kouphi	**Kou**	4	2	0
Amnisos	**Amn**	2	1	4	Krasi	**Kra**	2	2	5
Apodoulou	**Apd**	4	2	5	Kritsa	**Kri**	1	2	0
Apidi	**Api**	1	2	6	Lasithi	**Las**	2	2	0
Archanes	**Arh**	2	2	3, 4, 6,8	Ligortinos	**Lig**	3	2	6
Arkalochori	**Ark**	2	2	5, 7	Loutraki	**Lou**	4	2	4
Armenoi	**Arm**	4	2	6	Makryyialos	**Mak**	1	1	4
Artsa	**Art**	2	2	6	Malia	**Mal**	2	1	1, 2
Arvi	**Arv**	1	1	5	Martha	**Mar**	1	2	0
Axos	**Axo**	4	2	0	Mavrospilia	**Mas**	2	2	6
Berbani	**Ber**	0	0	0	Melidoni	**Mel**	2	2	7
Chania	**Cha**	4	1	3, 6	Mesara	**Mes**	3	2	0
Chersonisos	**Che**	2	1	0	Milatos	**Mil**	1	1	6, 0
Crete	**Cre**	0	0	0	Mochlos	**Moc**	1	1	3, 6
Dreros	**Dre**	1	2	6	Mohos	**Moh**	2	2	0
Episkopi / Ped.	**EkP**	2	2	6	Moires	**Moi**	3	2	6
Elakhanes	**Elk**	0	0	0	Mouliana	**Mol**	1	1	6
Elounda	**Elo**	1	1	0	Moni	**Mon**	4	2	6
Eparchia Pedias	**EpP**	2	2	0	Moria	**Mor**	3	2	6
Galia	**Gal**	3	2	6	Mouri	**Mou**	2	1	0
Gazi	**Gaz**	2	1	6	Megali Vrisi	**Mvr**	3	2	0
Goulas	**Gol**	0	0	0	Myrsini	**Myr**	1	1	6
Gonies	**Gon**	4	2	5	Neo Chorio	**NeC**	3	2	6
Gournes	**Gor**	2	2	6	New Hospital	**NeH**	2	2	6
Gournia	**Gou**	1	1	3	Nerokouro	**Nek**	4	2	4
Graditsa	**Gra**	0	0	0	Nirou Chani	**NiC**	2	1	4
Grivigla	**Grv**	4	2	8	Olous	**Olo**	1	1	6
Gypsades	**Gyp**	2	2	6	Pakhyammos	**Paa**	1	1	6
Hagios Giorgos	**HaG**	1	2	6	Palaikastro	**Pak**	1	1	3, 6
Hagios Ioannis	**HaI**	2	2	6	Patso	**Pat**	4	2	7
Hagios Syllas	**HaS**	2	2	6	Pano Zakros	**PaZ**	1	2	0
Hagia Triadha	**HaT**	3	2	4, 6	Petsofas	**Pet**	1	2	8, 6
Helenes	**Hel**	4	2	5	Phaistos	**Pha**	3	2	1, 2
Heraklion	**Her**	2	1	0	Pharmakokephalo	**Phk**	1	2	6
Idean Cave	**IdC**	4	2	7	Phodele	**Pho**	4	2	0
Ierapetra	**Iep**	1	1	0	Pigi	**Pig**	4	2	6
Isopata	**Iso**	2	2	6	Piskokephalo	**Pik**	1	2	0
Itanos	**Ita**	1	1	4	Pines	**Pin**	1	2	0
Iuktas	**Iuk**	2	2	8	Plakoures	**Pla**	0	0	0
Kalo Chorio	**KaC**	1	2	5	Poros-Katsambas	**Pok**	2	1	3, 6
Kato Episkopi	**KaE**	1	2	6	Pompia	**Pom**	3	2	6
Kamilaris	**Kal**	3	2	6	Praisos	**Pra**	1	2	6
Kamares	**Kam**	3	2	5, 7	Pseira	**Pse**	1	1	3
Kardamoutsa	**Kar**	0	0	0	Psychro	**Psy**	1	2	7
Kasanoi	**Kas**	2	2	6	Pyrgiotissa	**Pyr**	2	0	0
Katsamba	**Kat**	2	1	3, 6	Rethymno	**Ret**	4	1	0
Kalamvaki	**Kav**	1	2	0	Roussaias Ekklisia	**RoE**	1	2	0
Keratokambos	**Kek**	1	1	0	Rogdia	**Rog**	4	2	0
Kephala	**Kep**	2	2	6	Routasi	**Rou**	3	2	5
Kharakas	**Kha**	0	0	0	Samonas	**Sam**	4	2	5
Kalochorafitis	**Khf**	3	2	6	Sellopoulo	**Sel**	3	2	6
Khonos	**Kho**	1	2	0	Sitia	**Sit**	1	1	0
Kalami	**Kla**	4	1	6	Skalia	**Ska**	1	2	5

Abbreviations and coding of sites, cont.

Site	Abbr.	G	L	S
Sphaka	**Sph**	1	2	6
Stavromenos	**Sta**	1	2	4
Stavrochori	**Stc**	1	2	0
Sternes	**Ste**	1	2	4
Stamnioi	**Stm**	2	2	6
Styrana	**Sty**	0	0	0
Sybrita	**Syb**	4	2	5
Sykologos	**Syk**	1	2	0
Syme	**Sym**	1	2	8
Tefeli	**Tef**	2	2	6
Tourloti	**Tou**	1	1	4
Trochali Location	**Tro**	1	1	0
Tylissos	**Tyl**	4	2	4
Vai	**Vai**	1	1	4
Vavari	**Vav**	0	0	0
Vorou	**Vor**	0	0	0
Vryses	**Vrs**	1	2	6
Vrysinas	**Vry**	4	2	8
Zakros	**Zak**	1	1	1, 2
Zapher Papoura	**ZaP**	2	2	6
Ziros	**Zir**	1	2	5

Coding of sites

G = Geographical area
  0 = Unknown
  1 = Eastern Crete
  2 = North Central Crete
  3 = South Central Crete
  4 = Western Crete

L = Geographical location
  0 = Unknown
  1 = Coastal
  2 = Inland

G = Site type
  0 = Unknown
  1 = Palace
  2 = Palace surrounding
  3 = Town
  4 = Villa
  5 = Village
  6 = Tomb
  7 = Cave
  8 = Other cult place
  9 = Other

Appendix II.3

Museum abbreviations. Total 30.

AshM	Ashmolean Museum, Oxford
AthM	Nat. Archaeological Museum, Athens
BarC	Collection Barbier, Geneva
BerM	Antikenmuseen, Berlin
BosM	Museum of Fine Arts, Boston
BriM	British Museum, London
BSA	British School at Athens
CaM	Fogg art Museum, Cambridge, USA
CaUM	Cambridge University Museum
FiwM	Fitzwilliam Museum, Cambridge
GiaC	Giamalakis Collection, Herakleion
HerM	Archaeological Museum, Herakleion
HNM	Archaeological Museum, Hagios Nikolaos
IerM	Archaeological Museum, Ierapetra
Kass	Staatl. Kunstsamlung Antikenabt., Kassel
LeiM	Rijksmuseum van Oudheden, Leiden
LonM	Museum of London
Louv	Louvre, Paris
MaM	Collection Borély, Marseilles
MeC	Metaxa Collection, Herakleion
MeM	Metropolitan Museum of Arts, New York
OrtC	Collection G. Ortiz, Geneva
PhiM	Philadelphia Museum, USA
PigM	Museo Preistorico, Luigi Pigorino, Rome
RetM	Archaeological Museum, Rethymno
RoM	Museo Nazionale di Villa Giulia, Rome
SaOM	Museum Saint-Omer
SciC	Scientific Collection, Herakleion
StraM	Stratigraphical Museum, Knossos
WieM	Kunsthistorishes Museum, Wien

Only museum number refers to Herakleion Museum

Appendix III.   Copper ingots

Appendix III / 1

Appendix III. Copper ingots from Crete sorted by ingot type / condition / period / site. Total 75

No	Site	Ct	G	L	S	Pe	Da	Ot	C	Le	Wi	Th	We	Cu	Sn	Pb	As	Sb	Fe	Ni	Co	Zn	Bi	Tot	Pr	Mu	Ref
30	Pak	0		1	3	0	0		1	40	37	0	0	0	0	0	0	0	0	0	0	0	0	0	0	HerM	Evely, 00, 344, no 35
29	Pak	0	1	1	3	0	0		1	39	34	0	0	88.5	0	0	0	0	0	0.2	0	0	0	88.7	0	HerM	Evely, 00, 344, no 34; Mangou / Ioannou, 00, 213
18	HaT	Vano 7	3	2	4	1	4	1	1	0	0	0	0	86.8	0	0	0.1	0	0.1	0	0	0.8	0.1	87.8	5	726I	Evely, 00, 343; Mangou / Ioannou, 00, 213
5	HaT	Vano 7	3	2	4	1	4	1	1	37	33	5.5	29.5	82.4	0	0.1	0.1	0	0.2	0	0	1	0.1	84.4	5	725	Evely, 00, 343, no 10; Mangou / Ioannou, 00, 213
4	HaT	Vano 7	3	2	4	1	4	1	1	35	34	0	0	86.9	0	0	0.1	0	0.1	0	0	0.8	0.1	87.9	5	725	Evely, 00, 343, no 9; Mangou / Ioannou, 00, 213
13	HaT	Vano 7	3	2	4	1	4	1	1	0	0	0	0	0	0	0	0	0	0	0	0	0	0	0	5	HerM	Mangou / Ioannou, 00, 213
19	HaT	Vano 7	3	2	4	1	4	1	1	0	0	0	0	0	0	0	0	0	0	0	0	0	0	0	5	PigM	Evely, 00, 343; Mangou / Ioannou, 00, 213
3	HaT	Vano 7	3	2	4	1	4	1	1	37	35	0	0	84.4	0	0	0.1	0	0	0	0	2	0.1	86.7	5	724	Evely, 00, 343, no 8; Mangou / Ioannou, 00, 213
8	HaT	Vano 7	3	2	4	1	4	1	1	0	0	0	0	87.2	0	0	0	0	0	0	0	0.6	0.1	87.8	5	HerM	Mangou / Ioannou, 00, 213
1	HaT	Vano 7	3	2	4	1	4	1	1	45	39	4	27.3	82	0	0	0.6	0	0.7	0.6	0.4	0.5	0.1	85.1	5	721	Evely, 00, 343, no 6; Mangou / Ioannou, 00, 213
10	HaT	Vano 7	3	2	4	1	4	1	1	0	0	0	0	88.1	0	0	0.1	0	0.1	0	0	0.1	0.1	88.5	5	HerM	Mangou / Ioannou, 00, 213
2	HaT	Vano 7	3	2	4	1	4	1	1	37	35	6	27	87.3	0	0	0.1	0	0	0	0	0.4	0	87.9	5	722	Evely, 00, 343, no 7; Mangou / Ioannou, 00, 213
7	HaT	Vano 7	3	2	4	1	4	1	1	0	0	0	0	88.3	0	0	0.1	0	0.1	0	0	0.1	0.1	88.5	5	HerM	Mangou / Ioannou, 00, 213
6	HaT	Vano 7	3	2	4	1	4	1	1	0	0	0	0	87	0	0	0.1	0	0.1	0	0	0.4	0.1	87.6	5	HerM	Mangou / Ioannou, 00, 213
12	HaT	Vano 7	3	2	4	1	4	1	1	0	0	0	0	86.2	0	0	0.6	0	0	0.1	0	0.9	0.1	87.8	5	HerM	Mangou / Ioannou, 00, 213
11	HaT	Vano 7	3	2	4	1	4	1	1	0	0	0	0	87.3	0	0	0	0	0.9	0	0.1	0.4	0.1	88.9	5	HerM	Mangou / Ioannou, 00, 213
16	HaT	Vano 7	3	2	4	1	4	1	1	0	0	0	0	88.1	0	0	0	0	0	0	0	0.3	0.1	88.6	5	HerM	Mangou / Ioannou, 00, 213
15	HaT	Vano 7	3	2	4	1	4	1	1	0	0	0	0	85.8	0	0	0.1	0	0.1	0	0.1	0.7	0.1	86.8	5	HerM	Mangou / Ioannou, 00, 213
14	HaT	Vano 7	3	2	4	1	4	1	1	0	0	0	0	83.8	0	0.1	0.1	0	0.3	0	0.1	1.1	0.1	85.6	5	HerM	Mangou / Ioannou, 00, 213
9	HaT	Vano 7	3	2	4	1	4	1	1	0	0	0	0	86	0	0.1	0.1	0	4.2	0	0.2	0.3	0.1	91	5	726b	Mangou / Ioannou, 00, 213
31	PoK	Tryp. Hill	2	1	3	0	0	1	1	0	0	0	0	0	0	0	0	0	0	0	0	0	0	0	5	0	Dimopoulou, 97, 435, Plate CLXXIa
28	Tyl	House A	4	2	4	1	4	1	1	0	0	0	0	87.1	0	0	0	0	0.1	0	0	0.8	0.1	88.3	5	1763b	Mangou / Ioannou, 00, 213
27	Tyl	House A	4	2	4	1	4	1	1	0	0	0	0	87.1	0	0	0.2	0.2	0	0	0	0.1	0	87.6	5	1763a	Mangou / Ioannou, 00, 213
26	Tyl	House A	4	1	4	1	4	1	1	35	35	7.5	26.5	86.5	0	0	0.1	0	3.1	0.2	0	0.2	0.1	90.1	5	1763	Evely, 00, 344, no 25; Mangou / Ioannou, 00, 213
25	Zak	XI	1	1	1	1	4	1	1	38	32	7	29	88	0	0.2	0.2	0	0.2	0	0	0.2	0.1	88.8	5	2606	Platon, 88, II, 226; Mangou / Ioannou, 00, 213
21	Zak	XI	1	1	1	1	4	1	1	41	32	5	29	88.2	0	0.1	0.5	0	0.9	0.1	0.3	0.2	0.1	90.4	5	2602	Platon, 88, II, 225; Mangou / Ioannou, 00, 213
22	Zak	XI	1	1	1	1	4	1	1	37	30	6.5	29	90.5	0	0	0.3	0	0.6	0.1	0.2	0.1	0.1	91.8	5	2603	Platon, 88, II, 225; Mangou / Ioannou, 00, 213
23	Zak	XI	1	1	1	1	4	1	1	45	42	3.5	29	90	0	0	0	0	0	0	0	0	0.1	90.2	5	2604	Platon, 88, II, 226; Mangou / Ioannou, 00, 213
20	Zak	XI	1	1	1	1	4	1	1	39	30	6.3	29	87.8	0	0	0.3	0	0	0	0	0	0.1	88.4	5	2601	Platon, 88, II, 226; Mangou / Ioannou, 00, 213
24	Zak	XI	1	1	1	1	4	1	1	37	34	6.2	29	86.9	0	0	0	0	0	0	0	0	0.1	87	5	2605	Platon, 88, II, 225; Mangou / Ioannou, 00, 213
60	Kom	0	3	1	3	0	0	1	3	4	3	2.5	0.04	0	0	0	0	0	0	0	0	0	0	0	0	M 3	Blitzer, 95, 501, no M 3
59	Kom	0	3	1	3	0	0	1	3	4	3.2	2	0.05	0	0	0	0	0	0	0	0	0	0	0	0	M 2	Blitzer, 95, 501, no M 2
62	Kom	0	3	1	3	0	0	1	3	4.6	3.7	1.6	0.08	0	0	0	0	0	0	0	0	0	0	0	3	M 5	Blitzer, 95, 501, no M 5
63	Kom	0	3	1	3	0	0	1	3	5	3.5	2.3	0.14	0	0	0	0	0	0	0	0	0	0	0	3	M 6	Blitzer, 95, 501, no M 6
55	Cha	0	4	1	3	0	4	1	3	0	0	0	0	0	0	0	0	0	0	0	0	0	0	0	3	0	Stos-Gale et.al. 00, 207, no 4
34	Gou	0	1	1	3	1	4	1	3	0	0	0	0	0	0	0	0	0	0	0	0	0	0	0	3	B	Betancourt, et al, 78, 7; Stos-Gale et al, 97, 110
33	Gou	0	1	1	3	1	4	1	3	0	0	0	0	96.1	0	0	0	0	0	0	0	0	0.1	96.5	3	D	Bet.et al, 78, 7; M./ I., 00, 213; S-G.et al,97, 110
35	Gou	0	1	1	3	1	4	1	3	0	0	0	0	0	0	0	0	0	0	0	0	0	0	0	3	C	Betancourt et al, 78, 7; Stos-Gale et al, 97, 110
32	Gou	0	1	1	3	1	4	1	3	0	0	0	0	87.4	0	0	0.2	0	3.8	0	0.1	0	0	91.5	3	A	Bet.et al, 78, 7; M. / I., 00, 213; S-G. et al, 97,110
17	HaT	Vano 7	3	2	4	1	4	1	1	0	0	0	0	87	0	0	0.4	0	0	0	0.1	0	0.1	87.7	5	HerM	Mangou / Ioannou, 00, 213
36	Kno	0	2	2	1	1	0	1	3	0	0	0	0	88.6	0	0	0.2	0	0	0	0.1	0	0	89	5	1962	Evely, 00, 344, no 33; Mangou / Ioannou, 00, 213
47	Moc	Build. A	1	1	3	1	4	1	3	0	0	0	0	0	0	0	0	0	0	0	0	0	0	0	3	0	Soles / Davaras, 94, 416, 01, 77
51	Moc	Build. A	1	1	3	1	4	1	3	0	0	0	0	0	0	0	0	0	0	0	0	0	0	0	3	0	Soles / Davaras, 94, 416, 01, 77
54	Moc	Build. A	1	1	3	1	4	1	3	0	0	0	0	0	0	0	0	0	0	0	0	0	0	0	3	0	Soles / Davaras, 94, 419, 01, 77
50	Moc	Build. A	1	1	3	1	4	1	3	0	0	0	0	0	0	0	0	0	0	0	0	0	0	0	3	0	Soles / Davaras, 94, 416, 01, 77
49	Moc	Build. A	1	1	3	1	4	1	3	0	0	0	0	0	0	0	0	0	0	0	0	0	0	0	3	0	Soles / Davaras, 94, 416, 01, 77
52	Moc	Build. A	1	1	3	1	4	1	3	0	0	0	0	0	0	0	0	0	0	0	0	0	0	0	3	0	Soles / Davaras, 94, 416, 01, 77
39	Moc	House C	1	1	3	1	4	1	3	0	0	0	0	0	0	0	0	0	0	0	0	0	0	0	3	0	Soles / Davaras, 96, 201, 01, 77
38	Moc	House C	1	1	3	1	4	1	3	0	0	0	0	0	0	0	0	0	0	0	0	0	0	0	3	0	Soles / Davaras, 96, 201, 01, 77
41	Moc	Build. A	1	1	3	1	4	1	3	0	0	0	0	0	0	0	0	0	0	0	0	0	0	0	3	0	Soles / Davaras, 94, 415, 01, 77
40	Moc	Build. A	1	1	3	1	4	1	3	0	0	0	0	0	0	0	0	0	0	0	0	0	0	0	3	0	Soles / Davaras, 94, 415, 01, 77
46	Moc	Build. A	1	1	3	1	4	1	3	0	0	0	0	0	0	0	0	0	0	0	0	0	0	0	3	0	Soles / Davaras, 94, 416, 01, 77

Appendix III / 2

Appendix III. Copper ingots from Crete sorted by ingot type / condition / period / site. Total 75

No	Site	Ct	G	L	S	Pe	Da	Ot	C	Le	Wi	Th	We	Cu	Sn	Pb	As	Sb	Fe	Ni	Co	Zn	Bi	Tot	Pr	Mu	Ref
53	Moc	Build. A	1	1	3	1	4	1	3	0	0	0	0	0	0	0	0	0	0	0	0	0	0	0	3	0	Soles / Davaras, 94, 416, 01, 77
42	Moc	Build. A	1	1	3	1	4	1	3	0	0	0	0	0	0	0	0	0	0	0	0	0	0	0	3	0	Soles / Davaras, 94, 416, 01, 77
45	Moc	Build. A	1	1	3	1	4	1	3	0	0	0	0	0	0	0	0	0	0	0	0	0	0	0	3	0	Soles / Davaras, 94, 416, 01, 77
43	Moc	Build. A	1	1	3	1	4	1	3	0	0	0	0	0	0	0	0	0	0	0	0	0	0	0	3	0	Soles / Davaras, 94, 416, 01, 77
48	Moc	Build. A	1	1	3	1	4	1	3	0	0	0	0	0	0	0	0	0	0	0	0	0	0	0	3	0	Soles / Davaras, 94, 416, 01, 77
44	Moc	Build. A	1	1	3	1	4	1	3	0	0	0	0	0	0	0	0	0	0	0	0	0	0	0	3	0	Soles / Davaras, 94, 416, 01, 77
73	PoK	Sand. Plot	2	1	3	1	0	1	3	0	0	0	0	0	0	0	0	0	0	0	0	0	0	0	0	0	Dimopoulou, 1997, 435
74	PoK	Sand. Plot	2	1	3	1	0	1	3	0	0	0	0	0	0	0	0	0	0	0	0	0	0	0	0	0	Dimopoulou, 1997, 435
75	PoK	Sand. Plot	2	1	3	1	0	1	3	0	0	0	0	0	0	0	0	0	0	0	0	0	0	0	0	0	Dimopoulou, 1997, 435
70	PoK	Sand. Plot	2	1	3	1	0	1	3	0	0	0	0	0	0	0	0	0	0	0	0	0	0	0	0	0	Dimopoulou, 1997, 435
71	PoK	Sand. Plot	2	1	3	1	0	1	3	0	0	0	0	0	0	0	0	0	0	0	0	0	0	0	0	0	Dimopoulou, 1997, 435
72	PoK	Sand. Plot	2	1	3	1	0	1	3	0	0	0	0	0	0	0	0	0	0	0	0	0	0	0	0	0	Dimopoulou, 1997, 435
37	Zak	0	2	1	2	1	0	1	3	0	0	0	0	0	0	0	0	0	0	0	0	0	0	0	0	0	Evely, 00, 344, no 32
56	Cha	0	4	1	3	2	5	1	3	0	0	0	0	0	0	0	0	0	0	0	0	0	0	0	3	0	Stos-Gale et al, 00, 207, no 5
57	Cha	0	4	1	3	2	6	1	3	0	0	0	0	0	0	0	0	0	0	0	0	0	0	0	3	0	Stos-Gale et al; 00, 207, no 8
61	Kom	0	3	1	3	3	7	1	3	3	2.2	1.3	0.02	0	0	0	0	0	0	0	0	0	0	0	0	0	Blitzer, 95, 501, no M 4
58	Kom	0	3	1	3	3	0	1	3	3	2.8	1.6	0	0	0	0	0	0	0	0	0	0	0	0	0	0	Blitzer, 95, 501, no M 1
69	NiC	0	2	1	4	0	0	2	1	0	0	0	0	87.2	0.1	0	0.3	0.1	2.2	0.4	0.3	1.7	0.1	92.3	0	2407	Mangou / Ioannou, 00, 213
64	Ark	0	2	2	7	1	0	2	1	24	24	5.5	8	83.8	0	0	0.1	0	0.8	3.2	0.8	0.1	0.1	89.1	0	2407	Evely, 00, 343, no 1; Mangou / Ioannou, 00, 213
65	Ark	0	2	2	7	1	0	2	1	0	0	0	0	91.6	0.1	0	0.6	0.1	0.1	0.1	0	0	0.1	92.8	0	2408	Evely, 00, 343, no 2; Mangou / Ioannou, 00, 213
68	Ark	0	2	2	7	1	0	2	1	0	0	0	0	88.6	0.1	0	0.1	0.1	1.2	0	0	0.6	0.1	90.9	0	2411	Evely, 00, 343, no 5; Mangou / Ioannou, 00, 213
67	Ark	0	2	2	7	1	0	2	1	0	0	0	0	87.2	0.1	0	0.1	0.1	0.8	0	0	0.8	0.1	89.2	0	2410	Evely, 00, 343, no 4; Mangou / Ioannou, 00, 213
66	Ark	0	2	2	7	1	0	2	1	0	0	0	0	89	0.1	0	0.1	0.1	0.4	0	0	0.4	0.1	90.1	0	2409	Evely, 00, 343, no 3; Mangou / Ioannou, 00, 213

Appendix IV.   Metallurgical objects

Appendix IV.1

Crucibles

Bowl on low pierced stem

No	Provenience	Dating	Material	Diameter	Height, bowl	Height, tot.	Condition	Mus. no	Reference
1	Aghia Photia	EM I - II	Clay	-	-	-	Intact	-	Evely, 2000, 347, no 1
2	Aghia Photia	EM I - II	Clay	-	-	-	Intact	-	Evely, 2000, 347, no 2
3	Myrtos Pyrgos	MM II	Clay	12	6	10	4 fragments	-	Evely, 2000, 347, no 3
4	Knossos, Hou. Mon.Pil.	MM IIIA	Clay	10	2 (fragm)	4+ ?	2 fragments	-	Evely, 2000, 347, no 3a
5	Knossos, R/R	?	Clay	12-9		8	Fragments	-	Evely, 2000, 347, no 4
6	Gournia	MM III - LM I	Clay	7		14	Intact	-	Evely, 2000, 347, no 5
7	Kommos, Civic Centre	MM III	Clay	30 (est)			Fragment	-	Blitzer, 1995, 502, no M 7
8	Kommos, Civic Centre	LM I	Clay				Rim fragment	-	Blitzer, 1995, 503, no M 8
9	Kommos, Civic Centre	MM III - LM IA	Clay	24 (est)			2 fragments	-	Blitzer, 1995, 503, no M 9
10	Kommos, Civic Centre	MM III - LM IA	Clay				Bowl fragment	-	Blitzer, 1995, 503, no M 10
11	Kommos, Civic Centre	MM III - LM IA	Clay				Bowl fragment	-	Blitzer, 1995, 503, no M 11
12	Kommos, Civic Centre	MM - LM IA / B	Clay				Bowl fragment	-	Blitzer, 1995, 503, no M 12
13	Kommos, Civic Centre	MM III? - LM I	Clay				2 spout fragments	-	Blitzer, 1995, 503, no M 13
14	Kommos, Civic Centre	MM III? - LM I	Clay				Bowl fragment	-	Blitzer, 1995, 503, no M 14
15	Kommos, Civic Centre	MM III - LM I	Clay				Rim fragment	-	Blitzer, 1995, 503, no M 15
16	Kommos, Civic Centre	MM III	Clay				Rim fragment	-	Blitzer, 1995, 503, no M 16
17	Kommos, Civic Centre	MM III - LM IB?	Clay				Bowl fragment	-	Blitzer, 1995, 504, no M 17

Hemispherical, clay

No	Provenience	Dating	Material	Length	Width	Height	Condition	Mus. no	Reference
18	Malia, Qu. Mu	MM II	Clay	12	6.6	4.2	Intact	-	Poursat, 1996, 116, no M 71/28
19	Knossos, UM P	LM II	Clay	9.8	8.8	4.7	Intact	-	Evely, 2000, 347, no 6
20	Knossos, UM P	LM II	Clay	9.7	9	3.5	Nearly intact	-	Evely, 2000, 347, no 7
21	Knossos, UM H	LM II	Clay	6.5	6	3.5	Intact	-	Evely, 2000, 347, no 8
22	Knossos, UM D	LM II	Clay	9.1	6.2	5	Nearly intact	-	Evely, 2000, 347, no 9
23	Knossos, UM N	LM II	Clay	-	-	-	Bridge fragment	-	Evely, 2000, 347, no 10
24	Knossos, UM H	LM II	Clay	21	20	8	Fragmentary	-	Evely, 2000, 347, no 11
25	Knossos, UM H	LM II	Clay	6.3	4.3	2.7	Nearly intact	-	Evely, 2000, 347, no 12
26	Knossos, UM H	LM II	Clay	7 +	5.5	2.9	Nearly intact	-	Evely, 2000, 349, no 13
27	Knossos, UM M	LM II	Clay	6.2 +	5.8	2.5	Nearly intact	-	Evely, 2000, 349, no 14
28	Knossos, UM M	LM II	Clay	8.5	7.5	3.9	Nearly intact	-	Evely, 2000, 349, no 15
29	Knossos, UM M	LM II	Clay	8 (est.)	5	3	3 fragments	-	Evely, 2000, 349, no 16
30	Knossos, UM P	LM II	Clay	6	5.5	3.1	Nearly intact	-	Evely, 2000, 349, no 17
31	Knossos, UM N	LM II	Clay	5	6	-	Fragments	-	Evely, 2000, 349, no 18
32	Knossos, UM H	LM II	Clay	-	3.5 +	2	Fragments	-	Evely, 2000, 349, no 19

No	Provenience	Dating	Material	Length	Width	Height	Condition	Mus.no	Reference
33	Knossos, UM M	LM II	Clay	4+	-	4.5	Fragments	-	Evely, 2000, 349, no 20
34	Knossos, UM M	LM II	Clay	-	-	-	5 fragments	-	Evely, 2000, 349, no 21
35	Knossos, UM M	LM II	Clay	5+	4+	-	Fragments	-	Evely, 2000, 349, no 22
36	Knossos, UM L	LM II	Clay	5+	3+	-	Fragments	-	Evely, 2000, 349, no 23
37	Knossos, UM L	LM II	Clay	12+	12+	4	Fragments	-	Evely, 2000, 349, no 24
38	Knossos, UM L	LM II	Clay	7.5+	7.5+	4	2 fragments	-	Evely, 2000, 349, no 25
39	Knossos, UM N	LM II	Clay	7.5+	7.5+	-	Fragments	-	Evely, 2000, 349, no 26
40	Knossos, UM	LM II	Clay	10	10	-	4 fragments	-	Evely, 2000, 349, no 27
41	Knossos, UM	LM II	Clay	14+	-	8+	Fragment	-	Evely, 2000, 349, no 28
42	Kommos, Hill Top/Side	LM III?	Clay	-	-	-	Bowl fragment	-	Blitzer, 1995, 504, no M 18
43	Kommos, Hill Top/Side	LM IIIA2?	Clay	15	15	5.3	Fragments	-	Blitzer, 1995, 504, no M 19
44	Kommos, Hill Top/Side	LM IIIA2-IIIB	Clay	14(est)	14(est)	6.9	Fragments	-	Blitzer, 1995, 504, no M 20
45	Kommos, Hill Top/Side	LM IIIB	Clay	15	15	8	8 fragments	-	Blitzer, 1995, 505, no M 21
46	Kommos, Hill Top/Side	LM IIIA ?	Clay	-	-	-	Body fragment	-	Blitzer, 1995, 505, no M 22
47	Kommos, Hill Top/Side	LM	Clay	-	-	-	Body fragment	-	Blitzer, 1995, 505, no M 23
48	Kommos, Hill Top/Side	LM IIIB	Clay	-	-	-	Body Fragment	-	Blitzer, 1995, 505, no M 24
49	Kommos, Hill Top/Side	LM IIIA1/2?	Clay	-	-	-	3 fragments	-	Blitzer, 1995, 505, no M 25
50	Kommos, Hill Top/Side	LM IIIA-IIIB	Clay	-	-	-	5 fragments	-	Blitzer, 1995, 505, no M 26
51	Kommos, Hill Top/Side	LM IIIB ?	Clay	-	-	-	Rim fragment	-	Blitzer, 1995, 505, no M 27
52	Kommos, Hill Top/Side	LM IIIA?	Clay	-	-	-	Body fragment	-	Blitzer, 1995, 505, no M 28
53	Palaikastro	LM IIIB	Clay	21	22	11	13 fragments	-	Hemingway, 1996, 221, no 3427
54	Palaikastro	LM IIIB	Clay	15+	-	-	7 fragments	-	Hemingway, 1996, 223, no 3428
55	Palaikastro	LM IIIB	Clay	-	-	-	6 fragments	-	Hemingway, 1996, 223, no 2429

Stone

No	Provenience	Dating	Material	Length	Width	Height	Condition	Mus.no	Reference
56	Gournia, C 24	LM I	Steatite	13	13	7	Intact	317a	Evely, 2000, 349, no 33

Unspecified

No	Provenience	Dating	Material	Length	Width	Height	Condition	Mus.no	Reference
57	Knossos, Strat. Mus.	MM	Clay	-	-	-	Fragments	-	Evely, 2000, 351, no 39
58	Knossos, Little Palace	MM III - LM I ?	Clay	-	-	-	Fragments	-	Evely, 2000, 351, no 40
59	Knossos, R/R	LM I ?	Clay	-	-	-	Fragments	-	Evely, 2000, 351, no 45
60	Knossos, Hog. House	LM I/II	Clay	-	-	-	Fragment	-	Evely, 2000, 351, no 46
61	Knossos, Strat. Mus.	?	Clay	-	-	-	Fragment	-	Evely, 2000, 351, no 48
62	Gypsades, Tomb XVIII	MM III	Clay	-	-	-	Fragment	-	Evely, 2000, 351, no 38
63	Zakros, Hou. of Niches	LM I	Clay	-	-	-	Fragment?	-	Evely, 2000, 351, no 41
64	Chania	LM IB	Clay	-	-	-	Fragment?	-	Stos-Gale et.al., 2000, 207
65	Chania	LM IIIB2	Clay	-	-	-	Fragment?	-	Stos-Gale et.al., 2000, 207
66	Chania	LM IIIB2	Clay	-	-	-	Fragment?	-	Stos-Gale et.al., 2000, 207

Appendix IV.2

Moulds
Open stone moulds

No	Provenience	Dating	Material	No of moulds	Moulds for	Object dimens.	Condition	Mus. no	Reference
1	Malia, Qu Mu	MM II	Schist	3+1	3 chisels, 1 pick	L. 19, 33, 19; 26	Intact	-	Poursat, 1996, 55, no C 23
2	Malia, Palace	MM III	Talc schist	1	Circular object	Dia. 20	Fragment	-	Evely, 2000, 356, no 8
3	Malia, Palace	MM III	Talc schist	2	2 circular objects	Dia. 17, 26	Fragment	2244	Evely, 2000, 356, no 7
4	Malia, Palace	MM III	Talc schist	1	Ovoid object	Dia. 16+	Fragment	3227	Evely, 2000, 356, no 6
5	Malia, Palace	MM III	Talc schist	1+1+1+1	3 bars, 1 kite	L. 10, 9, 11, 8	Intact	2190	Evely, 2000, 356, no 5
6	Malia, Palace	MM III	Talc schist	1	Bar	L. 5	Intact	2245	Evely, 2000, 358, no 13
7	Malia, Palace	MM III	Talc schist	2	Kite, ?	L. 8	Fragment	2191	Evely, 2000, 358, no 14
8	Malia, House E	MM III-LM I	Schist	1+1(double)	Dagger	L. 13	Intact	2473	Evely, 2000, 360, no 25
9	Gournia, Fh	LM I	Stone	1	Axe	L. 3	Fragment	398	Evely, 2000, 358, no 10
10	Gournia, Fh	LM I	Stone	1	Knife	L.13	Intact	339	Evely, 2000, 356, no 1
11	Gournia, Fh	LM I	Stone	1	Nail?	L.6	Intact	400	Evely, 2000, 356, no 2
12	Gournia, E 10	LM I	Schist	1+5+4+3	4 chisels, 9 bars	L.23-28, 6-11	Repaired	397	Evely, 2000, 356, no 3
13	Gournia, Fh	LM I	Stone	1	Rod / rivet ?	L. 5	Fragments	401	Evely, 2000, 361, no 35
14	Phaistos	MM III?	Limestone	1	Chisel?	L.32	Intact	234	Evely, 2000, 358, no 11
15	Phaistos, R/ 69	MM III?	Stone	1	Curved blade	L.12	Intact	233	Evely, 2000, 358, no 12
16	Phaistos, R/ 71	MM III-LM I	Schist	1+1+1	2 chisels, 1 rod	L. 8, 6, 15	Broken	232	Evely, 2000, 358, no 15
17	Phaistos	MM III?	Limestone	1+1	Hammer?	L. 9	Fragmentary	231	Evely, 2000, 360, no 27
18	Palaikastro	?	Stone	2+1	3 bars	L. 9, ??	?	-	Evely, 2000, 358, no 16
19	Mochlos. Buil.A	LM I	Stone	1	Nail?	-	Intact	-	Soles / Davaras, 1994, 416

Two-piece stone moulds

No	Provenience	Dating	Material	No/ moulds	Mould for	Object dimensios	Condition	Mus. no	Reference
20	Koumasa	EM-MM	Stone	1	Double-axe	-	Intact	-	Evely, 2000, 358, no 18
21	Malia, Qu. Mu	MM II	Schist	1	Double-axe	-	Fragment	-	Poursat, 1996, 116, no M89/2202
22	Malia, Qu. Mu	MM II	Schist	1	Double-axe	L. 15	Fragment	-	Poursat, 1996, 116, no M 78/C33
23	Malia, Qu. Mu	MM II	Schist	1	Double-axe	-	Fragment	-	Poursat, 1996, 116, no C 18
24	Malia, Qu, Mu	MM II	Schist	1	Tool	-	Fragment	-	Poursat, 1996, 55, no C 27
25	Malia, Qu. Mu	MM II	Schist	1	Pick	-	Fragment	-	Poursat, 1996, 116, no 56 S 547
26	Myrtos Pyrgos	MM II-III	Schist	1	Spear?	-	Fragment	-	Evely, 2000, 360, no 26
27	Malia, NW area	MM III	Talc schist	1	Double-axe	L. 15	Intact	2187	Evely, 2000, 358, no 19
28	Malia, NW area	MM III	Talc schist	1	Double-axe	L. 14	Fragments	2186	Evely, 2000, 358, no 20
29	Malia, NW area	MM III	Talc schist	1	Double-axe	L. 14	Fragment	2189	Evely, 2000, 360, no 23
30	Malia, NW area	MM III	Talc schist	1	Double-axe	L. 14	Fragment	2188	Evely, 2000, 360, no 22
31	Malia, NW area	MM III	Talc schist	1	Double-axe	L. 14	Fragment	2184	Evely, 2000,360, no 24
32	Malia, NW area	MM III	Talc schist	1	Double-axe	L. 15	Fragment	2185	Evely, 2000,358, no 21

No	Provenience	Dating	Material		Mould for	Object dimens.	Condition	Mus. no	Reference
33	Malia, Hou. E	MM III-LM I	Schist	1+1 (single)	Double-axe	L. 14	Intact	2473	Evely, 2000, 360, no 25
34	Malia	MM-LM?	Schist	1	?	-	One part	-	Evely, 2000, 360, no 28
35	Phaistos	MM III ?	Limestone	1+1(single)	Double-axe	L. 10	One part	231	Evely, 2000, 360, no 27
36	Zakros, Hou/Nich.	LM I	Steatite	1	Double-axe ?	-	?	-	Platon, 1988, 141

Clay investment moulds

No	Provenience	Dating	Material		Mould for	Object dimens.	Condition	Mus. no	Reference
37	Mochlos, Bui.A	LM IB	Clay		Handle	Dia. 2.3	Intact	-	Soles / Davaras, 1994, 416
38	Mochlos, Bui.A	LM IB	Clay		Handle	Dia. 2.3	One half	-	Soles / Davaras, 1994, 416
39	Palaikastro, Bui.I	LM II	Clay		Double-axe	-	One half	-	Evely, 2000, 360, no 29a
40	Kommos	LM IIIA2-IIIB	Clay, 2 layers		Double-axe	-	2 fragments	-	Blitzer, 1995, 506, no M 29
41	Kommos	LM IIIA2-IIIB	Clay, 2 layers		Double-axe	L.19	50 fragments	-	Blitzer, 1995, 506, no M 30
42	Kommos	LM IIIB	Clay, 2 layers		Double-axe	-	19 fragments	-	Blitzer, 1995, 506, no M 31
43	Kommos	LM IIIA	Clay, 2 layers		Double-axe	-	Fragment	-	Blitzer, 1995, 507, no M 32
44	Kommos	LM IIIA-IIIB	Clay, 2 layers		Double-axe	-	6 fragments	-	Blitzer, 1995, 507, no M 33
45	Kommos	LM IIIA2	Clay, 2 layers		Double-axe	-	Fragment	-	Blitzer, 1995, 507, no M 34
46	Kommos	LM IIIB ?	Clay, 2 layers		Double-axe	-	Fragment	-	Blitzer, 1995, 507, no M 35
47	Kommos	LM IIIB1	Clay, 2 layers		Double-axe	-	Fragment	-	Blitzer, 1995, 507, no M 36
48	Kommos	LM IIIB ?	Clay, 2, layers		Double-axe	-	Fragment	-	Blitzer, 1995, 507, no M 37
49	Kommos	LM IIIA-IIIB	Clay, 2 layers		Double-axe	-	Fragment	-	Blitzer, 1995, 507, no M 38
50	Kommos	LM IIIB1	Clay, 2 layers		Double-axe	-	Fragment	-	Blitzer, 1995, 507, no M 39
51	Kommos	LM IIIA-IIIB	Clay, 2 layers		Double-axe	-	Fragment	-	Blitzer, 1995, 507, no M 40
52	Palaikastro	LM IIIB	Clay, 2 layers		Double-axe	L. 17	8 fragments	-	Hemingway, 1996, 226, no 3437
53	Palaikastro	LM IIIB	Clay, 2 layers		Double-axe	-	2 fragments	-	Hemingway, 1996, 226, no 3438
54	Palaikastro	LM IIIB	Clay, 2 layers		Double-axe	-	Fragment	-	Hemingway, 1996, 226, no 3439
55	Palaikastro	LM IIIB	Clay, 2 layers		Axe	-	Fragment	-	Hemingway, 1996, 228, no 4630
56	Palaikastro	LM IIIB	Clay, 2 layers		Axe	-	Fragment	-	Hemingway, 1996, 228, no 4631
57	Palaikastro	LM IIIB	Clay, 2 layers		Axe	-	Corner piece	-	Hemingway, 1996, 228, no 4633
58	Palaikastro	LM IIIB	Clay, 2 layers		Sickle ?	-	Body fragment	-	Hemingway, 1996, 228, no 4639
59	Palaikastro	LM IIIB	Clay, 2 layers		Bars	-	Several fragments	-	Hemingway, 1996, 228, no 3440
60	Palaikastro	LM IIIB	Clay, 2 layers		Handles	-	11 fragments / 7 moulds?	-	Hemingway, 1996, 228, no 3447
61	Palaikastro	LM IIIB	Clay, 2 layers		Handles	-	5 fragments / 4 moulds ?	-	Hemingway, 1996, 230, no 3445
62	Palaikastro	LM IIIB	Clay, 2 layers		Rods	-	11 fragments	-	Hemingway, 1996, 230, no 3446
63	Palaikastro	LM IIIB	Clay, 2 layers		Spiral form /stand?	-	3 fragments / 3 m0ulds?	-	Hemingway, 1996, 233, no 3442
64	Palaikastro	LM IIIB	Clay, 2 layers		Relief dek. /stand?	-	3 fragments	-	Hemingway, 1996, 233, no 3444

Other moulds

No	Provenience	Dating	Material		Mould for	Object dimens.	Condition	Mus. no	Reference
65	Vasiliki	EM II	Kopper / Bronze		Double-axe	L. 16	Half mould	1466	Evely, 2000, 358, no 17
66	Phaistos	MM I-II	Clay/ lost wax		Fist, natural size	-	Intact?	-	Evely, 2000, 360, no 29
67	Knossos, UM, L	LM II	Clay/ 2 faces		2 billets	L. 4.2; 4.5	Intact	-	Evely, 2000, 356, no 4
68	Pseira	LM IIIA2-IIIB	Schist		-	-	-	-	Evely, 2000, 361, no 36
69	Amnissos	LM III	-		Spike / spear butt	-	-	-	Evely, 2000, 361, no 37

Appendix IV.3

Tuyéres

No	Provenience	Dating	Material	Length	Width	Height	Condition	Reference
1	Malia, Qua. Mu	MM II	Clay	7	5.3	-	Fragment	Poursat, 1996, 117, no 69 M 1392
2	Malia, Qua. Mu	MM II	Clay	6.8	3.5	-	Fragment	Poursat. 1996, 117, no 69 M 1455
3	Malia, Qua. Mu	MM II	Clay	7	4	-	Fragment	Poursat, 1996, 117, no 70 M 632
4	Malia, Qua. Mu	MM II	Clay	5.8	4	-	Fragment	Poursat, 1996, 117, no 71 M 1885
5	Palaikastro	LM IIIB	Clay	11	6.2	15	Fragment	Hemingway, 1996, 218, no 3430
6	Palaikastro	LM IIIB	Clay	9	6	9.5	Fragment	Hemingway, 1996, 218, no 3431
7	Palaikastro	LM IIIB	Clay	9.5	6	8	Fragment	Hemingway, 1996, 218, no 3432
8	Palaikastro	LM IIIB	Clay	9.5	6	8	Fragment	Hemingway, 1996, 218, no 3433
9	Palaikastro	LM IIIB	Clay	5.5	4	2	Fragment	Hemingway, 1996, 218, no 3434
10	Palaikastro	LM IIIB	Clay	6	6.5	8	Fragment	Hemingway, 1996, 218, no 3435
11	Palaikastro	LM IIIB	Clay	6	2	6.8	Fragment	Hemingway, 1996, 220, no 3436
12	Palaikastro	LM IIIB	Clay	6	4.8	2.1	Fragment	Hemingway, 1996, 220, no 4605
13	Palaikastro	LM IIIB	Clay	8.4	6.5	2.5	Fragment	Hemingway, 1996, 220, no 4606
14	Palaikastro	LM IIIB	Clay	5.5	5.8	2.5	Fragment	Hemingway, 1996, 220, no 4607
15	Palaikastro	LM IIIB	Clay	6.5	3.8	2	Fragment	Hemingway, 1996, 220, no 4608
16	Palaikastro	LM IIIB	Clay	6.3	5.4	2.4	Fragment	Hemingway, 1996, 220, no 4609
17	Palaikastro	LM IIIB	Clay	6.4	5.9	1.8	Fragment	Hemingway, 1996, 220, no 4610
18	Palaikastro	LM IIIB	Clay	4.6	2.3	2.2	Fragment	Hemingway, 1996, 220, no 4611

Appendix IV.4

Pot bellows

No	Provenience	Dating	Material	Diameter (bowl)	Height (bowl)	Length (nozzel)	Condition	Reference
1	Chrysokamino	EM III	Clay	27	-	-	Fragments/ 10 bellows	Betancourt et.al. 1999, 359
2	Kommos	LM	Clay	35	14	34	Nearly intact	Blitzer,1995, 508,no M 42
3	Kommos	LM III ?	Clay	-	-	60(est)	Nozzle fragment	Blitzer,1995, 508,no M 43
4	Kommos	LM IIIA2 - IIIB1?	Clay	-	-	10 +	3 nozzle fragments	Blitzer,1995, 509,no M 44

Note. All dimensions in cm

Appendix V. Bronze objects

Appendix V.I Chemical analyses of bronze objects and metallic workshop evidence sorted by object type / period / site. Total 302

No	Site	Pe	Da	Ot	Cu	Sn	As	Pb	Met	AnDa	Mus	Ref	DBno
56	0	0	0	1	94	6	0	0	0	1964	0	Evely, 93, 11, no 177	267
39	HaT	0	0	1	81.4	11.2	0.4	0.4	AAS	1998	1231	Mangou / Ioannou, 1998, 94	0
38	HaT	0	0	1	88.7	1.2	2.4	0.1	AAS	1998	1229	Mangou / Ioannou, 1998, 94	0
11	Kno	0	0	1	80.5	13.6	0.5	0.1	AAS	1998	1448	Mangou / Ioannou, 1998, 94	0
55	Pal	0	0	1	83.9	8.6	0.4	0.3	AAS	1998	4554	Mangou / Ioannou, 1998, 94	0
54	Pal	0	0	1	85.7	9	0.4	0.3	AAS	1998	4553	Mangou / Ioannou, 1998, 94	0
52	Pal	0	0	1	93	1.7	1	0.2	AAS	1998	1392	Mangou / Ioannou, 1998, 94	0
48	Pal	0	0	1	87.9	5.3	0.4	0.1	AAS	1998	1137	Mangou / Ioannou, 1998, 94	0
57	Psy	0	0	1	90	9	0.4	0	0	1995	AshM	Evely, 93, 8, no 76; Northover / Evely, 95, 87	235
199	Seh	0	0	1	96.1	0.8	1	0	0	1995	AshM	Evely, 93, 10, no 100; Northover / Evely, 95, 87, no2, 95	1571
198	Seh	0	0	1	91.6	2.8	2.1	0.1	0	1995	AshM	Evely, 93, 10, no102; Northover / Evely, .95, 85, no1, 95	1570
58	Syb	0	0	1	93	7	0	0	0	0	AshM	Evely, 93, 8, no 52	251
263	Kno	2	5	1	91.4	8.6	0	0	XRF	1977	0	Popham, 84, 59, no M 29; Catling / Jones 77,62-63	1855
146	Kno	2	5	1	92	8	0	0	XRF	1977	0	Evely, 93, 10, no 137	175
144	Kno	2	5	1	91	9	0	0	XRF	1977	0	Evely, 93, 6, no 21	167
267	Kno	2	5	1	91	9	0	0	XRF	1977	0	Popham, 84, 60, no M 47; Catling / Jones,77, 62-63	0
142	Kno	2	5	1	91	9	0	0	XRF	1977	0	Evely, 93, 6, no 29	168
253	Kno	2	5	1	92.3	7.7	0	0	XRF	1977	0	Popham, 84, 51, no L 129; Catling / Jones, 77,62-63	0
261	Kno	2	5	1	89.6	10.4	0	0	XRF	1977	0	Popham, 84, 57, no M 9; Catling / Jones, 77, 62-63	1853
143	Kno	2	5	1	90	10	0	0	XRF	1977	0	Evely, 93, 10, no 94	173
145	Kno	2	5	1	90	10	0	0	XRF	1977	0	Evely, 93, 6, no 20	166
264	Kno	2	5	1	90.3	9.7	0	0	XRF	1977	0	Popham, 84, 59, no M 30; Catling / Jones, 77,62-63	1856
203	Cha	3	7	1	89	10.2	0.2	0	XRF	2000	ChaM	Stos-Gale et al, 00, 213, Table 2, 214,Table.3, no 16	1707
9	Kno	0	0	2	84.6	9.7	0.3	0.1	AAS	1998	1114	Mangou / Ioannou, 1998, 94	0
10	Kno	0	0	2	79.4	8.7	0.2	0.1	AAS	1998	1120	Mangou / Ioannou, 1998, 94	0
25	Pha	0	0	2	85.3	7.3	0.6	0.1	AAS	1998	352	Mangou / Ioannou, 1998, 94	0
26	Pha	0	0	2	90.4	5.7	0.4	0.8	AAS	1998	353	Mangou / Ioannou, 1998, 94	0
185	Kno	2	5	2	88	12	0	0	XRF	1977	0	Popham, 84, 29, no H 93	1482
178	Psy	3	0	2	94.8	4.6	0.3	0.3	0	1950	AshM	Boardman, 61, 19, no 67, 160	1303
181	Psy	3	0	2	91	8.2	0.4	0.2	0	1950	AshM	Boardman, 61, 19, no 70, 160	1306
179	Psy	3	0	2	89	8.2	1.3	0	0	1950	AshM	Boardman, 61, 19, no 68, 160	1304
177	Psy	3	0	2	91.2	5.7	0.4	2	0	1950	AshM	Boardman, 61, 19, no 63, 160	1299
169	Psy	3	0	2	91.2	5.7	0.4	2	0	1950	AshM	Boardman, 61, 19, no 63, 160	1299

Appendix V.I Chemical analyses of bronze objects and metallic workshop evidence sorted by object type / period / site. Total 302

Appendix V.1 / 2

No	Site	Pe	Da	Ot	Cu	Sn	As	Pb	Met	AnDa	Mus	Ref	DBno
180	Psy	3	0	2	88	10.9	0.3	0.7	0	1950	AshM	Boardman, 61, 19, no 69, 160	1305
67	0	0	0	3	99	1	0	0	0	1964	0	Evely, 93, 49, no 199	407
35	HaT	0	0	3	87.2	6	0.4	0	AAS	1998	1226	Mangou / Ioannou, 1998, 94	0
37	HaT	0	0	3	90.4	1.2	0.7	0.1	AAS	1998	1228	Mangou / Ioannou, 1998, 94	0
44	HaT	0	0	3	89.7	4	0.3	0.1	AAS	1998	1532	Mangou / Ioannou, 1998, 94	0
36	HaT	0	0	3	84.1	6.9	0.1	0.1	AAS	1998	1227	Mangou / Ioannou, 1998, 94	0
31	HaT	0	0	3	82.8	10.6	0.2	0.1	AAS	1998	830	Mangou / Ioannou, 1998, 94	0
13	Kno	0	0	3	88.9	6.4	0.2	0.2	AAS	1998	1802	Mangou / Ioannou, 1998, 94	0
18	Kno	0	0	3	89.1	3	0.7	0	AAS	1998	2078	Mangou / Ioannou, 1998, 94	0
61	Kno	0	0	3	91	6.4	1.4	1.5	0	1995	AshM	Evely, 93, 46, no 67; Northover / Evely, 95, 87, no 5	335
60	Pak	0	0	3	85.9	1.9	0.4	1	AAS	1998	1377	Evely, 93, 49, no 145c; Mangou / Ioannou, 98, 95	363
66	Pak	0	0	3	82.8	10.9	0.5	0.2	AAS	1998	1382	Evely, 93, 49, no 145e; Mangou / Ioannou, 98, 95	365
65	Pak	0	0	3	82.9	10	0.2	0.1	AAS	1998	1380	Evely, 93, 49, no 145d; Mangou / Ioannou, 98, 95	364
59	Pak	0	0	3	83.5	7.3	0.1	0.2	AAS	1998	1383	Evely, 93, 47, no 120; Mangou / Ioannou, 98, 95	357
64	Psy	0	0	3	91	7	1.1	0.4	0	1995	AshM	Evely, 93, 42, no 23; Northover / Evely, 95, 87, no6	390
62	Psy	0	0	3	88	11	0	0	0	1910	1380	Evely, 93, 42, no 25	392
63	Tou	0	0	3	99	0	0	0	0	1910	535	Evely, 93, 46, no72	398
88	0	1	0	3	95	2.3	2	0.2	AAS	1976	AshM	Evely, 93, 44, no 56	368
89	Kar	1	0	3	95	4.6	0	0.2	0	1956	BriM	Evely, 93, 42, no 14	416
87	Kno	1	0	3	95	5	0	0	XRF	1977	0	Evely, 93, 42, no 26	333
93	Kno	1	0	3	84.8	9.6	0.3	0.2	AAS	1998	2076	Evely, 93, 42, no4; Mangou / Ioannou, 98, 94	328
85	Kno	1	0	3	97	1.8	1.4	0.6	0	0	AshM	Evely, 93, 44, no 42	334
86	Kno	1	0	3	72.7	6.4	0.1	0	AAS	1998	2077	Evely, 93, 42, no5; Mangou / Ioannou, 98, 94	329
94	Nek	1	0	3	85	15	0.3	0	0	1984	ChaM	Evely, 93, 42, no 18	408
92	Nek	1	0	3	85	14	0.2	0	0	1984	0	Evely, 93, 42, no 19	409
77	Pak	1	0	3	85.7	6.3	0.5	0	AAS	1998	1379	Evely, 93, 47, no 136; Mangou / Ioannou, 98, 95	358
84	Pak	1	4	3	86.4	5.5	0.3	0.2	AAS	1998	1378	Evely, 93, 49, no 142; Mangou / Ioannou, 98, 95	359
91	Pak	1	4	3	91.9	4.7	0.4	0.1	AAS	1998	852	Evely, 93, 49, no 143; Mangou / Ioannou, 98, 94	360
78	Pak	1	4	3	87.9	0.2	1.1	0.1	AAS	1998	1376a	Evely, 93, 49, no 145a; Mangou / Ioannou, 98, 95	362
90	Pak	1	4	3	89	0.2	0.7	0.1	AAS	1998	1372	Evely, 93, 49, no 144; Mangou / Ioannou, 98, 95	361
74	Pha	1	4	3	88.2	6	0.2	0.2	AAS	1998	344	Evely, 93, 44, no 49; Mangou / Ioannou, 98, 94	317
80	Pha	1	4	3	96	1.6	0.5	0.3	AAS	1998	346	Evely, 93, 44, no 51; Mangou / Ioannou, 98, 94	319
79	Pha	1	4	3	88.2	6.7	0.1	0.1	AAS	1998	345	Evely, 93, 44, no 50; Mangou / Ioannou, 98, 94	318

Appendix V.I Chemical analyses of bronze objects and metallic workshop evidence sorted by object type / period / site. Total 302

No	Site	Pe	Da	Ot	Cu	Sn	As	Pb	Met	AnDa	Mus	Ref	DBno
75	Pha	1	4	3	89	5.2	0.2	0.1	AAS	1998	342	Evely, 93, 44, no 47; Mangou / Ioannou, 98, 94	315
81	Pha	1	4	3	89.6	3.7	0.2	0.1	AAS	1998	347	Evely, 93, 44, no 52; Mangou / Ioannou, 98, 94	320
76	Pha	1	4	3	91.4	3.6	0.5	0.2	AAS	1998	343	Evely, 93, 44, no 48; Mangou / Ioannou, 98, 94	316
82	Pha	1	4	3	97	0.2	0.6	0.1	AAS	1998	348	Evely, 93, 44, no 53; Mangou / Ioannou, 98, 94	321
83	Pha	1	4	3	95	2.6	0.4	0.2	AAS	1998	349	Evely, 93, 44, no 54; Mangou / Ioannou, 98, 94	322
200	Seh	1	0	3	91.9	6.8	0.5	0.1	0	1995	AshM	Evely, 93, 46, no 96; Northover / Evely, 95, 87, no4, 95	1573
147	Kno	2	5	3	99	1	0	0	XRF	1977	0	Evely, 93, 47, no 134	339
69	HaT	0	0	5	90.7	0.2	0.3	0	AAS	1998	1223	Evely, 93, 63, no 5; Mangou / Ioannou, 98, 94	468
20	Kno	0	0	5	88.2	0.2	0.2	0	AAS	1998	2080	Mangou / Ioannou, 1998, 94	0
70	Kno	0	0	5	98	0.7	0.2	0.6	0	0	AshM	Evely, 93, 63, no 13	475
19	Kno	0	0	5	92.5	0.2	0.2	0	AAS	1998	2079	Mangou / Ioannou, 1998, 94	0
68	Kno	0	0	5	99	0	0.5	0	0	1995	AshM	Evely, 93, 63, no 12; Northover / Evely, 95, 87, no7	474
12	Kno	0	0	5	91.8	0.2	0.2	0.1	AAS	1998	1800	Mangou / Ioannou, 1998, 94	0
51	Pal	0	0	5	83.3	8.7	5.6	0.1	AAS	1998	1384	Mangou / Ioannou, 1998, 94	0
28	Pha	0	0	5	87.4	6.3	0.3	2.3	AAS	1998	1038	Mangou / Ioannou, 1998, 94	0
71	Mil	0	0	6	91	7.9	0.8	0.4	0	1995	AshM	Evely, 93, 68, no 8; Northover / Evely, 95, 87, no 9	486
95	Moc	1	0	6	89	7.5	0	3.7	0	1995	AshM	Evely, 93, 68, no 6; Northover / Evely, 95, 87, no 6	481
96	Moc	1	0	6	87	6	0	6.9	0	1950	AshM	Evely, 93, 67, no 2	480
98	HaT	1	0	9	92	0.8	0.6	0	AAS	1998	701	Evely, 93, 31, no 23; Mangou / Ioannou, 98, 94	296
97	Kno	1	0	9	74.4	0.2	0.1	0.1	AAS	1998	2053	Evely, 93, 33, no 45; Mangou / Ioannou, 98, 94	288
286	Kno	2	5	9	79.6	20.4	0	0	XRF	1977	0	Popham, 84, 65, no M 139; Catling / Jones, 77, 62-63	0
148	Kno	2	5	9	80	20	0	0	XRF	1977	0	Evely, 93, 33, no 72	292
149	Kno	2	5	9	93	7	0	0	XRF	1977	0	Evely, 93, 31, no 42	287
45	HaT	0	0	11	85.8	9.3	0.5	0.1	AAS	1998	1533	Mangou / Ioannou, 1998, 94	0
32	HaT	0	0	11	84.5	5.6	0.3	0	AAS	1998	831	Mangou / Ioannou, 1998, 94	0
21	Kno	0	0	11	89.1	5.8	0.3	0.1	AAS	1998	2750	Mangou / Ioannou, 1998, 94	0
100	HaT	1	0	11	92.5	1.2	0.6	0	AAS	1998	1253	Evely, 93, 101, no 14; Mangou / Ioannou, 98, 94	569
99	Psy	1	0	11	77	14	1.2	6.9	0	1995	AshM	Evely, 93, 101, no 2; Northover / Evely, 95, 87, no 10	571
202	Cha	2	6	12	97	0.8	0.9	0	XRF	2000	ChaM	Stos-Gale et al, ,00, 213, Tabl. 2, 214, Tabl.3, .no 6	1706
151	Kno	2	5	12	90	9	0	1	XRF	1977	0	Evely, 93, 92, no 133	548
295	Kno	2	5	12	90	0	9	1	XRF	1977	0	Popham, 84, 78, no P 6; Catling / Jones, 77, 62-63	0
150	Kno	2	5	12	97	3	0	0	XRF	1977	0	Evely, 93, 92, no 96	536
152	Kno	2	5	12	91	9	0	0	XRF	1977	0	Evely, 93, 88, no 37	519

Appendix V.I Chemical analyses of bronze objects and metallic workshop evidence sorted by object type / period / site. Total 302

No	Site	Pe	Da	Ot	Cu	Sn	As	Pb	Met	AnDa	Mus	Ref	DBno
216	Kno	2	5	12	90.4	9.6	0	0	XRF	1977	0	Popham, 84, 23, no H 15; Catling / Jones, 77, 62-63	0
213	Kno	3	9	12	89.1	10.9	0	0	XRF	1977	0	Popham, 84, 16, no G 2; Catling / Jones, 77, 62-63	0
234	Kno	2	5	13	90.5	9.5	0	0	XRF	1977	0	Popham, 84, 31, no H 99; Catling / Jones, 77, 62-63	0
221	Kno	2	5	13	89.7	10.3	0	0	XRF	1977	0	Popham, 84, 24, no H 33; Catling / Jones, 77, 62-63	0
233	Kno	2	5	13	86.6	13.4	0	0	XRF	1977	0	Popham, 84, 29, no H 97; Catling / Jones, 77, 62-63	0
229	Kno	2	5	13	97	3	0	0	XRF	1977	0	Popham, 84, 29, no H 80; Catling / Jones, 77, 62-63	0
226	Kno	2	5	13	98.5	1.5	0	0	XRF	1977	0	Popham, 84, 26, no H 72; Catling / Jones, 77, 62-63	0
206	Cha	3	8	13	98	0.8	0.6	0	XRF	2000	ChaM	Stos-Gale et al, 00, 213, Tabl. 2, 214, Tabl.3, .no 24	1710
27	Pha	0	0	14	92	4.6	0.3	0.2	AAS	1998	364	Mangou / Ioannou, 1998, 94	0
297	Kno	2	5	14	96.7	3.3	0	0	XRF	1977	0	Popham, 84, 80, no P 9; Catling / Jones, 77, 62-63	0
242	Kno	2	5	14	98.5	1.5	0	0	XRF	1977	0	Popham, 84, 45, no L 24; Catling / Jones, 77, 62-63	0
301	Kno	2	5	14	98.5	1.5	0	0	XRF	1977	0	Popham, 84, 87, no P130; Catling / Jones, 77, 62-63	0
269	Kno	2	5	14	98.8	1.2	0	0	XRF	1977	0	Popham, 84, 60, no M 65; Catling / Jones, 77, 62-63	0
274	Kno	2	5	14	100	0	0	0	XRF	1977	0	Popham, 84, 62, no M 93; Catling / Jones, 77, 62-63	0
296	Kno	2	5	14	92.8	7.2	0	0	XRF	1977	0	Popham, 84, 80, no P 8; Catling / Jones, 77, 62-63	0
271	Kno	2	5	14	96.6	2.4	1	0	XRF	1977	0	Popham, 84, 61, no M 82; Catling / Jones, 77, 62-63	0
204	Cha	3	7	14	91	8.3	0.1	0	XRF	2000	ChaM	Stos-Gale et al, 00, 213, Tabl. 2, 214, Tabl.3, .no17	1708
186	Kno	2	5	15	99	1	0	0	XRF	1977	0	Popham, 84, 51, no L 126	1484
187	Kno	2	5	18	99	1	0	0	XRF	1977	0	Popham, 84, 51, no L 128	1485
196	Psy	3	0	18	88.8	10.9	0.3	0.1	0	1950	AshM	Boardman, 61, 51, no 221, 160	1566
40	HaT	0	0	20	86.4	5.8	0.6	2.8	AAS	1998	1232	Mangou / Ioannou, 1998, 94	0
41	HaT	0	0	20	91.6	0.6	0.3	0	AAS	1998	1242	Mangou / Ioannou, 1998, 94	0
14	Kno	0	0	20	83.1	9.3	0.1	0	AAS	1998	1804	Mangou / Ioannou, 1998, 94	0
50	Pal	0	0	20	89	0.2	0.7	0.1	AAS	1998	1372	Mangou / Ioannou, 1998, 94	0
53	Pal	0	0	20	91.5	0.2	0.5	0.1	AAS	1998	4552	Mangou / Ioannou, 1998, 94	0
220	Kno	2	5	20	87.8	12.2	0	0	XRF	1977	0	Popham, 84, 24, no H 32; Catling / Jones, 77, 62-63	0
228	Kno	2	5	20	90.4	9.6	0	0	XRF	1977	0	Popham, 84, 29, no H 77; Catling / Jones, 77, 62-63	0
217	Kno	2	5	20	90	10	0	0	XRF	1977	0	Popham, 84, 23, no H 16; Catling / Jones, 77, 62-63	0
33	HaT	0	0	21	91.5	0.8	0.7	0.9	AAS	1998	833a	Mangou / Ioannou, 1998, 94	0
43	HaT	0	0	21	90.4	0.4	0.1	0	AAS	1998	1254b	Mangou / Ioannou, 1998, 94	0
29	HaT	0	0	21	89.6	0.2	0.4	0.3	AAS	1998	703a	Mangou / Ioannou, 1998, 94	0
42	HaT	0	0	21	88.9	0.2	0.7	0.2	AAS	1998	1254a	Mangou / Ioannou, 1998, 94	0
47	HaT	0	0	21	91.9	0.2	0.9	1.1	AAS	1998	1544a	Mangou / Ioannou, 1998, 94	0

Appendix V.I Chemical analyses of bronze objects and metallic workshop evidence sorted by object type / period / site. Total 302

No	Site	Pe	Da	Ot	Cu	Sn	As	Pb	Met	AnDa	Mus	Ref	DBno
34	HaT	0	0	21	90.8	0.2	0.8	0	AAS	1998	833b	Mangou / Ioannou, 1998, 94	0
30	HaT	0	0	21	82.7	0.2	0.4	0.1	AAS	1998	703b	Mangou / Ioannou, 1998, 94	0
46	HaT	0	0	21	89.1	0.2	1	3.5	AAS	1998	1544	Mangou / Ioannou, 1998, 94	0
15	Kno	0	0	21	69.5	6.8	0.1	0	AAS	1998	2069	Mangou / Ioannou, 1998, 94	0
16	Kno	0	0	21	87.2	0.6	0.2	0.1	AAS	1998	2073a	Mangou / Ioannou, 1998, 94	0
5	Kno	0	0	21	84.4	0.2	0.3	1.7	AAS	1998	1089a	Mangou / Ioannou, 1998, 94	0
6	Kno	0	0	21	66.8	0.2	0.2	0.1	AAS	1998	1089b	Mangou / Ioannou, 1998, 94	0
3	Kno	0	0	21	64.7	4.3	0	0.1	AAS	1998	1088a	Mangou / Ioannou, 1998, 94	0
4	Kno	0	0	21	81.6	2.4	0.2	0.1	AAS	1998	1088b	Mangou / Ioannou, 1998, 94	0
7	Kno	0	0	21	82.9	10.3	0.3	0	AAS	1998	1092	Mangou / Ioannou, 1998, 94	0
2	Kno	0	0	21	81.1	9.7	0.1	0	AAS	1998	1087b	Mangou / Ioannou, 1998, 94	0
1	Kno	0	0	21	82.7	8.8	0.2	0	AAS	1998	1087a	Mangou / Ioannou, 1998, 94	0
17	Kno	0	0	21	82.7	0.9	0.3	0	AAS	1998	2073b	Mangou / Ioannou, 1998, 94	0
72	Pha	0	0	21	98	0	0	0	0	1920	HerM	Matthäus, 80, 104, no 63, 105	876
101	Kno	1	3	21	87.2	6.7	0.1	0	AAS	1998	2075	Matthäus, 80, 10, 261, no 370; Mangou / Ioannou, 98, 94	780
104	Mal	1	2	21	100	0	0	0	0	1980	2195	Matthäus, 80, 11, 125, no 117	786
103	Moc	1	4	21	100	0	0	0	0	1980	1581	Matthäus, 80, 123, no 112, 124	799
102	Tyl	1	4	21	100	0	0	0	0	1980	HerM	Matthäus, 80, 82, no1; Driessen / Macdonald, 97, 129	810
236	Kno	2	5	21	91.2	8.8	0	0	XRF	1977	0	Popham, 84, 37, no H 196; Catling / Jones, 77, 62-63	0
247	Kno	2	5	21	90.9	9.1	0	0	XRF	1977	0	Popham, 84, 47, no L 39; Catling / Jones, 77, 62-63	0
241	Kno	2	5	21	95.1	4.9	0	0	XRF	1977	0	Popham, 84, 45, no L 23; Catling / Jones, 77, 62-63	0
224	Kno	2	5	21	94.1	5.9	0	0	XRF	1977	0	Popham, 84, 24, no H 49; Catling / Jones, 77, 62-63	0
284	Kno	2	5	21	90.8	9.2	0	0	XRF	1977	0	Popham, 84, 64, no M 131; Catling / Jones, 77, 62-63	0
222	Kno	2	5	21	85.6	12.4	1	1	XRF	1977	0	Popham, 84, 24, no H 36; Catling / Jones, 77, 62-63	0
250	Kno	2	5	21	83.6	16.4	0	0	XRF	1977	0	Popham, 84, 48, no L 67; Catling / Jones, 77, 62-63	0
227	Kno	2	5	21	90.3	9.7	0	0	XRF	1977	0	Popham, 84, 29, no H 74; Catling / Jones, 77, 62-63	0
218	Kno	2	5	21	87.9	12.1	0	0	XRF	1977	0	Popham, 84, 24, no H 26; Catling / Jones, 77, 62-63	0
219	Kno	2	5	21	99.5	0.5	0	0	XRF	1977	0	Popham, 84, 24, no H 31; Catling / Jones, 77, 62-63	0
246	Kno	2	5	21	96	4	0	0	XRF	1977	0	Popham, 84, 47, no L 35; Catling / Jones, 77, 62-63	0
243	Kno	2	5	21	95.2	4.8	0	0	XRF	1977	0	Popham, 84, 45, no L 25; Catling / Jones, 77, 62-63	0
212	Sel	2	6	21	91.5	9.5	0	0	XRF	1976	0	Catling / Jones, 76, 22, no 37	0
209	Sel	2	6	21	80	20	0	0	XRF	1976	0	Catling / Jones, 76, 22, no 35	917
207	Sel	2	6	21	84.2	15.8	0	0	XRF	1976	0	Catling / Jones, 76, 22, no 19	904

Appendix V.I Chemical analyses of bronze objects and metallic workshop evidence sorted by object type / period / site. Total 302

Appendix V.1 / 6

No	Site	Pe	Da	Ot	Cu	Sn	As	Pb	Met	AnDa	Mus	Ref	DBno
210	Sel	2	6	21	85	15	0	0	XRF	1976	0	Catling / Jones, 76, 22, no 30	918
211	Sel	2	6	21	90.6	9.4	0	0	XRF	1976	0	Catling / Jones, 76, 22, no 36	921
208	Sel	2	6	21	93.4	6.6	0	0	XRF	1976	0	Catling / Jones, 76, 22, no 28	911
153	ZaP	2	6	21	88	9.1	0	1.1	0	0	AshM	Matthäus, 80, 42, 99, no 38	895
154	Sel	2	6	31	90	10	1	0	XRF	1976	HerM	Kilian-Dirlmeier, 93, 60, no 122; Catling / Jones, 76, 22	623
170	ZaP	3	0	31	82.9	11.1	0.3	0	AAS	1998	1099	Kilian-Dirlmeier, 93, 60, no 123; Mangou / Ioannou, 98, 94	609
171	ZaP	3	0	31	82.7	12.2	0.4	0.1	AAS	1998	1100	Kilian-Dirlmeier, 93, 59, no 114; Mangou / Ioannou, 98, 94	606
49	Pal	0	0	32	83.7	7.8	0.6	0.4	AAS	1998	1359	Mangou / Ioannou, 1998, 94	0
197	Pat	0	0	32	94.9	1.8	1.6	0.1	0	1950	AshM	Boardman, 61, 78, no 376, 160	1569
201	Gou	1	1	32	86.9	13.1	0	0	0	1968	CaUM	Charles, 68, 278	1645
298	Kno	2	5	32	91.6	8.4	0	0	XRF	1977	0	Popham, 84, 80, no P 10; Catling / Jones, 77, 62-63	0
22	Kno	0	0	33	86.4	7.9	0.8	0	AAS	1998	2972	Mangou / Ioannou, 1998, 94	0
8	Kno	0	0	33	91	10.8	0.3	0	AAS	1998	1110	Mangou / Ioannou, 1998, 94	0
23	Kno	0	0	33	81.1	12.1	0.4	0.1	AAS	1998	2974	Mangou / Ioannou, 1998, 94	0
24	Pha	0	0	33	88.7	5.5	0.5	0.2	AAS	1998	351	Mangou / Ioannou, 1998, 94	0
193	Psy	0	0	33	93.1	6.6	0	0.1	0	1950	AshM	Boardman, 61, 28, no 100, 160	1512
191	Psy	0	0	33	90.2	9.4	0.3	0	0	1950	AshM	Boardman, 61, 28, no 98, 160	1510
190	Psy	0	0	33	88.2	7.9	1.7	0.1	0	1950	AshM	Boardman, 61, 28, no 97, 160	1509
194	Psy	0	0	33	99	0.1	0	0.8	0	1950	AshM	Boardman, 61, 28, no 112, 160	1524
192	Psy	0	0	33	95.2	3.4	0.6	0.1	0	1950	AshM	Boardman, 61, 28, no 99, 160	1511
182	Kno	2	5	33	83	17	0	0	XRF	1977	0	Popham, 84, 26, no H 62	1477
183	Kno	2	5	33	89	11	0	0	XRF	1977	0	Popham, 84, 26, no H 63	1478
273	Kno	2	5	33	89.4	8.6	2	0	XRF	1977	0	Popham, 84, 61, no M 90; Catling / Jones, 7, 62-63	0
184	Kno	2	5	34	100	0	0	0	XRF	1977	0	Popham, 84, 26, no H 193	1479
205	Cha	3	8	34	94	5.8	0.5	0	XRF	2000	ChaM	Stos-Gale et al, 00, 213, Tabl. 2, 214, Tabl.3. no 23	1709
155	Sel	2	6	40	80	20	0	0	XRF	1976	0	Catling / Catling, 74, 230, no 12; Catling / Jones, 76, 22, no 8	1188
156	Sel	2	6	40	91.5	8.5	0	0	XRF	1976	0	Catling / Catling, 74, 229, no 11; Catling / Jones, 76, 22, no 6	1187
195	Psy	1	0	45	99.5	0.1	0.2	0.1	0	1950	AshM	Boardman, 61, 44, 45, no 203, 160	1543
73	Psy	0	0	46	98.5	0	0	0.6	AAS	1976	AshM	Sapouna-Sakellarakis, 95, 32, no 46, 153	977
124	0	1	0	46	94.8	0	0	4.9	AAS	1976	MetC	Sapouna-Sakellarakis, 95, 91, no 157, 153	1065
128	0	1	0	46	98	0.1	0	0	AAS	1976	LonM	Sapouna-Sakellarakis, 95, 93, no 162, 153	1070
126	0	1	0	46	96	3	0	1	AAS	1976	SaOM	Sapouna-Sakellarakis, 95, 94, no 164, 153	1072
125	0	1	0	46	95.4	1.3	1.2	0.1	AAS	1976	BerM	Sapouna-Sakellarakis, 95, 93, no 160, 153	1068

Appendix V.I Chemical analyses of bronze objects and metallic workshop evidence sorted by object type / period / site. Total 302

Appendix V.1 / 7

No	Site	Pe	Da	Ot	Cu	Sn	As	Pb	Met	AnDa	Mus	Ref	DBno
127	Arh	1	0	46	98.6	0	0	0.6	AAS	1995	2508	Sapouna-Sakellarakis, 95, 45, no 74, 157	1002
123	Gou	1	4	46	95	4	0.1	0.2	AAS	1995	612	Sapouna-Sakellarakis, 95, 14, no 10, 155, 159	943
111	HaT	1	0	46	88.2	10.9	0	0.3	AAS	1995	4783	Sapouna-Sakellarakis, 95, 76, no 131, 157	1052
113	HaT	1	0	46	52.7	0.2	0	46.2	AAS	1995	2054	Sapouna-Sakellarakis, 95, 74, no 125, 155	1046
110	HaT	1	0	46	86.9	0.1	0	12.7	AAS	1995	4782	Sapouna-Sakellarakis, 95, 76, no 130, 155	1051
112	HaT	1	0	46	87.6	11.6	0	0.3	AAS	1995	4781	Sapouna-Sakellarakis, 95, 75, no 129, 157	1050
109	Juh	1	0	46	72.5	0	0	25.9	AAS	1995	4273	Sapouna-Sakellarakis, 95, 50, no 85, 157	1009
106	Juh	1	0	46	98.6	0	1.2	0.2	AAS	1995	4912	Sapouna-Sakellarakis, 95, 51, no 89, 160	1013
105	Kat	1	0	46	95.2	1	0.8	1.9	AAS	1995	1829	Sapouna-Sakellarakis, 95, 55, no 97, 157	1018
108	Psy	1	0	46	76.3	0.4	0.2	23	AAS	1995	435	Sapouna-Sakellarakis, 95, 24, no 25, 155	957
118	Psy	1	0	46	98	1	0.1	0.1	AAS	1995	4780	Sapouna-Sakellarakis, 95, 29, no 37, 155	968
115	Psy	1	0	46	60.6	0	0.2	38.9	AAS	1995	4530	Sapouna-Sakellarakis, 95, 28, no 35, 155	966
114	Psy	1	0	46	99	0	0	0	AAS	1976	Louv	Sapouna-Sakellarakis, 95, 34, no 53, 153	984
107	Psy	1	0	46	100	0	0	0	AAS	1976	AshM	Sapouna-Sakellarakis, 95, 30, no 41, 153	972
120	Sko	1	3	46	89.4	1.6	0.6	7.9	AAS	1995	2575	Sapouna-Sakellarakis, 95, 40, no 64, 157	995
119	Sko	1	3	46	96.3	0.5	1.1	0.2	AAS	1995	2574	Sapouna-Sakellarakis, 95, 39, no 63, 158	994
117	Sko	1	3	46	98.1	0.2	1	0.1	AAS	1995	2573	Sapouna-Sakellarakis, 95, 39, no 62, 158	993
122	Tyl	1	0	46	99.1	0.1	0.5	0	AAS	1995	1831	Sapouna-Sakellarakis, 95, 60, no 101, 157	1022
121	Tyl	1	0	46	97	1.7	0.4	0.1	AAS	1995	1762	Sapouna-Sakellarakis, 95, 59, no 100, 156	1021
116	Tyl	1	0	46	97.7	1.5	0.2	0	AAS	1995	1832	Sapouna-Sakellarakis, 95, 61, no 102, 157	1023
163	Dre	2	5	46	88.3	8.5	1	0.1	AAS	1995	2398	Sapouna-Sakellarakis, 95, 15, no 11, 158	1079
162	HaT	2	5	46	70.9	0.5	0	27.2	AAS	1995	753	Sapouna-Sakellarakis, 95, 70, no 118, 157	1039
165	IdC	2	5	46	81.6	10.3	0.4	6.6	AAS	1995	1641	Sapouna-Sakellarakis, 95, 66, no 112, 158	1033
159	Kno	2	5	46	76.9	12.2	1.2	7.1	AAS	1995	704	Sapouna-Sakellarakis, 95, 52, no 90, 158	1014
160	Mal	2	5	46	73.4	2.1	0.7	23	AAS	1995	2958	Sapouna-Sakellarakis, 95, 37, no 60, 157	991
157	Pal	2	5	46	94.6	0.3	0.7	0.1	AAS	1995	1418	Sapouna-Sakellarakis, 95, 10, no 2, 158, 160	940
158	Psy	2	5	46	99.5	0	0.4	0.1	AAS	1976	AshM	Sapouna-Sakellarakis, 95, 34, no 50, 153	981
161	Psy	2	5	46	98.8	0	0.2	0	AAS	1995	4529	Sapouna-Sakellarakis, 95, 28, no 34, 155	965
164	Psy	2	5	46	97.2	0	1	0.2	AAS	1995	204	Sapouna-Sakellarakis, 95, 18, no 13, 155	945
172	Cre	3	0	46	98.1	0	0.3	0	AAS	1976	BerM	Sapouna-Sakellarakis, 95, 81, no 140, 153	1058
173	Psy	3	0	46	99.4	0	0.3	0	AAS	1995	430	Sapouna-Sakellarakis, 95, 23, no 21, 155	953
174	Psy	3	0	46	99	0.4	0.1	0.3	AAS	1976	AshM	Sapouna-Sakellarakis, 95, 31, no 42, 153	973
136	HaT	1	0	47	99.3	0.2	0.2	0.1	AAS	1995	759	Sapouna-Sakellarakis, 95, 71, no 121, 157	1042

Appendix V.I Chemical analyses of bronze objects and metallic workshop evidence sorted by object type / period / site. Total 302

Appendix V.1 / 8

No	Site	Pe	Da	Ot	Cu	Sn	As	Pb	Met	AnDa	Mus	Ref	DBno
135	HaT	1	0	47	99.3	0.2	0.1	0	AAS	1995	760	Sapouna-Sakellarakis, 95, 72, no 122, 155	1043
132	HaT	1	0	47	92.7	0.1	0.5	5.3	AAS	1995	1029	Sapouna-Sakellarakis, 95, 73, no 124, 157	1045
137	HaT	1	0	47	85.3	0.2	0.2	13.7	AAS	1995	4527	Sapouna-Sakellarakis, 95, 75, no 127, 157	1048
139	Pal	1	0	47	96.9	0	0.9	0.1	AAS	1995	1417	Sapouna-Sakellarakis, 95, 9, no 1, 158, 160	939
133	Psy	1	0	47	84.1	0.9	0.1	12.5	AAS	1995	437	Sapouna-Sakellarakis, 95, 25, no 26, 155	958
131	Psy	1	0	47	99	0.8	0.3	0.1	AAS	1976	AshM	Sapouna-Sakellarakis, 95, 31, no 43, 153	974
129	Psy	1	0	47	93	3	1.2	0.6	AAS	1976	AshM	Sapouna-Sakellarakis, 95, 30, no 40, 153	971
130	Psy	1	0	47	97.6	0	0.6	0.2	AAS	1995	429	Sapouna-Sakellarakis, 95, 22, no 20, 155	952
134	Psy	1	0	47	74.5	0	0	24.5	AAS	1976	AshM	Sapouna-Sakellarakis, 95, 32, no 44, 153	975
138	Psy	1	0	47	72.8	0.4	0.4	25.4	AAS	1995	4779	Sapouna-Sakellarakis, 95, 28, no 36, 155	967
166	HaT	2	5	47	97.2	1.6	0.3	0.5	AAS	1995	758	Sapouna-Sakellarakis, 95, 71, no 120, 155	1041
176	Psy	3	0	47	70.5	1.2	1	27	AAS	1976	AshM	Sapouna-Sakellarakis, 95, 32, no 45, 153	976
175	Psy	3	0	47	99	0	0.9	0.2	AAS	1976	AshM	Sapouna-Sakellarakis, 95, 33, no 48, 153	979
141	Psy	1	0	49	99.5	0	0.4	0.1	AAS	1976	AshM	Sapouna-Sakellarakis, 95, 33, no 49, 153	980
140	Ret	1	0	49	96.4	1.5	0.5	0.7	AAS	1976	LonM	Sapouna-Sakellarakis, 95, 79, no 138,153	1057
167	Sel	2	6	51	0	8.4	0	0	XRF	1976	0	Catling / Catling, 74, 230, no 17; Catling / Jones, 76, 22, no 11	1193
168	Sel	2	6	51	0	10.6	0	0	XRF	1976	0	Catling / Catling, 74, 230, no 13; Catling / Jones, 76, 22, no10	1189
188	Kno	2	5	55	92.5	7.5	0	0	XRF	1977	0	Popham, 84, 50, no L 120	1486
189	Kno	2	5	55	92	8	0	0	XRF	1977	0	Popham, 84, 76, no O 2	1487
277	Kno	2	5	60	100	0	0	0	XRF	1977	0	Popham, 84, 62, no M 98; Catling / Jones, 77, 62-63	0
275	Kno	2	5	60	94.5	5.5	0	0	XRF	1977	0	Popham, 84, 62, no M 96; Catling / Jones, 77, 62-63	0
280	Kno	2	5	61	91.1	8.9	0	0	XRF	1977	0	Popham, 84, 64, no M 113; Catling / Jones, 77, 62-63	0
276	Kno	2	5	61	99.3	0.7	0	0	XRF	1977	0	Popham, 84, 62, no M 97; Catling / Jones, 77, 62-63	0
281	Kno	2	5	61	95	5	0	0	XRF	1977	0	Popham, 84, 64, no M 127; Catling / Jones, 77, 62-63	0
270	Kno	2	5	61	98.7	1.3	0	0	XRF	1977	0	Popham, 84, 60, no M 70; Catling / Jones, 77, 62-63	0
214	Kno	2	5	61	80.2	19.8	0	0	XRF	1977	0	Popham, 84, 23, no H 8; Catling / Jones, 77, 62-63	0
223	Kno	2	5	61	98.5	1.5	0	0	XRF	1977	0	Popham, 84, 24, no H 39; Catling / Jones, 77, 62-63	0
268	Kno	2	5	61	92.9	7.1	0	0	XRF	1977	0	Popham, 84, 60, no M 51; Catling / Jones, 77, 62-63	0
265	Kno	2	5	61	88.2	11.8	0	0	XRF	1977	0	Popham, 84, 59, no M 36; Catling / Jones, 77, 62-63	0
279	Kno	2	5	61	93.2	6.8	0	0	XRF	1977	0	Popham, 84, 62, no M 106; Catling / Jones, 77, 62-63	0
254	Kno	2	5	61	93.2	6.8	0	0	XRF	1977	0	Popham, 84, 51, no L 137; Catling / Jones, 77, 62-63	0
266	Kno	2	5	61	86.1	13.9	0	0	XRF	1977	0	Popham, 84, 59, no M 39; Catling / Jones, 77,62-63	0
257	Kno	2	5	61	93.9	6.1	0	0	XRF	1977	0	Popham, 84, 57, no M 3; Catling / Jones, 77, 62-63	0

Appendix V.I Chemical analyses of bronze objects and metallic workshop evidence sorted by object type / period / site. Total 302

Appendix V.1 / 9

No	Site	Pe	Da	Ot	Cu	Sn	As	Pb	Met	AnDa	Mus	Ref	DBno
237	Kno	2	5	61	92.3	7.7	0	0	XRF	1977	0	Popham, 84, 37, no H 202; Catling / Jones,77,62-63	0
238	Kno	2	5	61	96.5	3.5	0	0	XRF	1977	0	Popham, 84, 37, no H 203; Catling / Jones,77,62-63	0
290	Kno	2	5	61	94.2	5.8	0	0	XRF	1977	0	Popham, 84, 67, no M 186; Catling / Jones,77,62-63	0
230	Kno	2	5	61	86.9	13.1	0	0	XRF	1977	0	Popham, 84, 29, no H 87; Catling / Jones, 77, 62-63	0
287	Kno	2	5	61	91.5	8.5	0	0	XRF	1977	0	Popham, 84, 67, no M 182; Catling / Jones, 77, 62-63	0
302	Kno	2	5	61	80	20	0	0	XRF	1977	0	Popham, 84, 87, no P 134; Catling / Jones, 77, 62-63	0
262	Kno	2	5	61	89.5	10.5	0	0	XRF	1977	0	Popham, 84, 59, no M 24; Catling / Jones, 77,62-63	0
285	Kno	2	5	61	88.7	11.3	0	0	XRF	1977	0	Popham, 84, 64, no M 132; Catling / Jones, 77, 62-63	0
272	Kno	2	5	61	89.5	10.5	0	0	XRF	1977	0	Popham, 84, 61, no M 86; Catling / Jones, 77,62-63	0
240	Kno	2	5	61	93.5	6.5	0	0	XRF	1977	0	Popham, 84, 45, no L 20; Catling / Jones, 77, 62-63	0
244	Kno	2	5	61	98.3	1.7	0	0	XRF	1977	0	Popham, 84, 45, no L 27; Catling / Jones, 77, 62-63	0
258	Kno	2	5	61	95.8	4.2	0	0	XRF	1977	0	Popham, 84, 57, no M 5; Catling / Jones, 77, 62-63	0
239	Kno	2	5	61	93.4	6.6	0	0	XRF	1977	0	Popham, 84, 45, no L 13; Catling / Jones, 77, 62-63	0
283	Kno	2	5	61	91.9	8.1	0	0	XRF	1977	0	Popham, 84, 64, no M 130; Catling / Jones, 77, 62-63	0
289	Kno	2	5	61	95.3	4.7	0	0	XRF	1977	0	Popham, 84, 67, no M 185; Catling / Jones, 77, 62-63	0
288	Kno	2	5	61	83.4	16.6	0	0	XRF	1977	0	Popham, 84, 67, no M 184; Catling / Jones, 77, 62-63	0
248	Kno	2	5	61	93.8	6.2	0	0	XRF	1977	0	Popham, 84, 47, no L 41; Catling / Jones, 77, 62-63	0
282	Kno	2	5	61	95.5	4.5	0	0	XRF	1977	0	Popham, 84, 64, no M 129; Catling / Jones, 77, 62-63	0
251	Kno	2	5	62	91.5	8.5	0	0	XRF	1977	0	Popham, 84, 49, no L 73; Catling / Jones, 77, 62-63	0
294	Kno	2	5	62	99.5	0.5	0	0	XRF	1977	0	Popham, 84, 76, no O 3; Catling / Jones, 77, 62-63	0
252	Kno	2	5	62	90.4	9.6	0	0	XRF	1977	0	Popham, 84, 49, no L 90; Catling / Jones, 77, 62-63	0
225	Kno	2	5	62	98.2	1.8	0	0	XRF	1977	0	Popham, 84, 26, no H 64; Catling / Jones, 77, 62-63	0
255	Kno	2	5	62	99.1	0.9	0	0	XRF	1977	0	Popham, 84, 51, no L 140; Catling / Jones, 77, 62-63	0
292	Kno	2	5	62	99	1	0	0	XRF	1977	0	Popham, 84, 73, no N 32; Catling / Jones, 77, 62-63	0
291	Kno	2	5	62	95.1	4.9	0	0	XRF	1977	0	Popham, 84, 71, no N 10; Catling / Jones, 77, 62-63	0
293	Kno	2	5	62	98.9	1.1	0	0	XRF	1977	0	Popham, 84, 76, no N 76; Catling / Jones,.77, 62-63	0
278	Kno	2	5	62	100	0	0	0	XRF	1977	0	Popham, 84, 62, no M 103; Catling / Jones, 77, 62-63	0
299	Kno	2	5	62	90.5	9.5	0	0	XRF	1977	0	Popham, 84, 80, no P 33; Catling / Jones, 77, 62-63	0
259	Kno	2	5	62	99.5	0.5	0	0	XRF	1977	0	Popham, 84, 57, no M 6; Catling / Jones, 77, 62-63	0
300	Kno	2	5	62	85.5	14.5	0	0	XRF	1977	0	Popham, 84, 80, no P 34; Catling / Jones, 77, 62-63	0
215	Kno	2	5	62	98.3	1.7	0	0	XRF	1977	0	Popham, 84, 23, no H 9; Catling / Jones, 77, 62-63	0
256	Kno	2	5	62	98	2	0	0	XRF	1977	0	Popham, 84, 51, no L 141; Catling / Jones, 77, 62-63	0
232	Kno	2	5	62	91.7	8.3	0	0	XRF	1977	0	Popham, 84, 29, no H 95; Catling / Jones, 77, 62-63	0

Appendix V.I Chemical analyses of bronze objects and metallic workshop evidence sorted by object type / period / site. Total 302

Appendix V.1 / 10

No	Site	Pe	Da	Ot	Cu	Sn	As	Pb	Met	AnDa	Mus	Ref	DBno
231	Kno	2	5	62	94.8	5.2	0	0	XRF	1977	0	Popham, 84, 29, no H 90; Catling / Jones, 77, 62-63	0
260	Kno	2	5	62	98.7	1.3	0	0	XRF	1977	0	Popham, 84, 57, no M 7; Catling / Jones, 77, 62-63	0
249	Kno	2	5	62	98.7	1.3	0	0	XRF	1977	0	Popham, 84, 47, no L 42; Catling / Jones, 77, 62-63	0
235	Kno	2	5	63	99.1	0.9	0	0	XRF	1977	0	Popham, 84, 31, no H 102; Catling / Jones, 77, 62-63	0
245	Kno	2	5	63	95.4	3.6	1	0	XRF	1977	0	Popham, 84, 45, no L 30; Catling / Jones, 77, 62-63	0

Appendix V.2 / 1

Appendix V.2 Bronze objects. Sorted by object type / period / site. Total 1857

No	Site	Co	G	L	S	Pe	Da	Ot	C	TM	TK	TP	Cn	Le	Wi	He	Rd	We	Cu	Sn	As	Pb	Mu	Ref
230	0	0	4	0	0	0	0	1	1	0	0	0	1	19	1.3	0	0	0	0	0	0	0	ChaM	Evely, 93, 10, no 115
229	0	0	4	0	0	0	0	1	1	0	0	0	1	24	1.1	0	0	0	0	0	0	0	ChaM	Evely, 93, 10, no 109
267	0	0	0	0	0	0	0	1	1	0	0	0	1	0	0	0	0	0	94	6	0	0	0	Evely, 93, 11, no 177
228	0	0	4	0	0	0	0	1	1	0	0	0	1	20	1	0	0	0	0	0	0	0	ChaM	Evely, 93, 10, no 108
266	0	0	0	0	0	0	0	1	1	0	0	0	1	0	0	0	0	0	0	0	0	0	GiaC	Evely, 93, 10, no 133
265	0	0	0	0	0	0	0	1	1	0	0	0	1	0	0	0	0	0	0	0	0	0	GiaC	Evely, 93, 10, no 132
234	0	0	4	0	0	0	0	1	1	0	0	0	1	17	2	0	0	0	0	0	0	0	ChaM	Evely, 93, 8, no 75
264	0	0	0	0	0	0	0	1	1	0	0	0	1	0	0	0	0	0	0	0	0	0	GiaC	Evely, 93, 10, no 131
255	Arh	0	2	2	4	0	0	1	1	0	0	0	1	0	0	0	0	0	0	0	0	0	2485	Evely, 93, 10, no 120
209	HaT	NW Qua	3	2	4	0	0	1	1	0	0	0	2	23	0	0	0	0	0	0	0	0	0	Evely, 93, 11, no 175
210	HaT	NW Qua	3	2	4	0	0	1	1	0	0	0	2	24	0	0	0	0	0	0	0	0	0	Evely, 93, 11, no 175
212	HaT	Ro 37	3	2	4	0	0	1	1	0	0	0	1	0	0	0	0	0	0	0	0	0	0	Evely, 93, 11, no 176
211	HaT	Ro 37	3	2	4	0	0	1	1	0	0	0	1	0	0	0	0	0	0	0	0	0	0	Evely, 93, 11, no 176
206	HaT	0	3	2	4	0	0	1	1	0	0	0	1	30	1.5	0	0	0	0	0	0	0	0	Evely, 93, 10, no 92
205	HaT	0	3	2	4	0	0	1	1	0	0	0	1	7.3	1.6	0	0	0	0	0	0	0	1538	Evely, 93, 8, no 47
221	HaT	0	3	2	4	0	0	1	1	0	0	0	1	0	0	0	0	0	0	0	0	0	1538	Evely, 93, 8, no 53
222	HaT	0	3	2	4	0	0	1	1	0	0	0	1	0	0	0	0	0	0	0	0	0	1537	Evely, 93, 8, no 54
207	HaT	0	3	2	4	0	0	1	1	0	0	0	1	29	1	0	0	0	0	0	0	0	0	Evely, 93, 10, no 93
208	HaT	0	3	2	4	0	0	1	1	0	0	0	1	24	2.5	0	0	0	0	0	0	0	0	Evely, 93, 11, no141
258	Kam	0	3	2	5	0	0	1	1	0	0	0	1	0	0	0	0	0	0	0	0	0	2584	Evely, 93, 10, no 125
269	Kat	0	2	1	3	0	0	1	1	0	0	0	1	22	0	0	0	0	0	0	0	0	1726	Evely, 93, 10, no 104
268	Kat	0	2	1	3	0	0	1	1	0	0	0	1	27	0	0	0	0	0	0	0	0	1727	Evely, 93, 10, no 103
179	Kno	Roy. Road	2	2	2	0	0	1	1	0	0	0	3	3.8	0.7	0	0	0	0	0	0	0	0	Evely, 93, 11, no 168
178	Kno	Roy. Road	2	2	2	0	0	1	1	0	0	0	3	2.8	0.7	0	0	0	0	0	0	0	0	Evely, 93, 11, no 167
169	Kno	0	2	2	0	0	0	1	1	0	0	0	1	17	0.7	0	0	0	0	0	0	0	StraM	Evely, 93, 6,no 31
174	Kno	0	2	2	2	0	0	1	1	0	0	0	1	34	1.1	0	0	550	0	0	0	0	StraM	Evely, 93, 10, no 110
171	Kno	0	2	2	2	0	0	1	1	0	0	0	1	19	2.5	0	0	350	0	0	0	0	StraM	Evely, 93, 8,no 77
176	Kno	Roy. Road	2	2	2	0	0	1	1	0	0	0	1	3.8	0.9	0	0	0	0	0	0	0	0	Evely, 93, 11, no 146
181	Mal	Quart. Z	2	1	2	0	1	1	1	0	0	0	1	5.4	0.6	0	0	0	0	0	0	0	0	Evely, 93, 6, no 13
239	Mvr	0	3	2	0	0	0	1	1	0	0	0	2	25	1.1	0	0	0	0	0	0	0	506	Evely, 93, 10, no 98
238	Mvr	0	3	2	0	0	0	1	1	0	0	0	1	25	1.1	0	0	0	0	0	0	0	507	Evely, 93, 10, no 97
240	Mvr	0	3	2	0	0	0	1	1	0	0	0	1	30	1	0	0	0	0	0	0	0	505	Evely, 93, 10, no 99
160	Pak	Trial P	1	1	3	0	0	1	1	0	0	0	1	0	0	0	0	0	0	0	0	0	0	Evely, 93, 11, no 113
253	Pla	0	0	0	0	0	0	1	1	0	0	0	1	0	0	0	0	0	0	0	0	0	2730	Evely, 93, 10, no 124
252	Pla	0	0	0	0	0	0	1	1	0	0	0	1	0	0	0	0	0	0	0	0	0	2732	Evely, 93, 8, no 55
254	Pra	0	1	2	6	0	0	1	1	0	0	0	1	31	0	0	0	0	0	0	0	0	925	Evely, 93, 10, no 107
237	Psy	0	1	2	7	0	0	1	1	0	0	0	1	12	0	0	0	0	0	0	0	0	491	Evely, 93, 11, no 165
235	Psy	0	1	2	7	0	0	1	1	0	0	0	2	19	1.9	0	0	275	90	9	0.4	0	AshM	Evely, 93, 8, no 76; Northover / Evely, 95, 87
236	Psy	0	1	2	7	0	0	1	1	0	0	0	1	11	0.7	0	0	0	0	0	0	0	488	Evely, 93, 11, no 142

Appendix V.2 Bronze objects. Sorted by object type / period / site. Total 1857

No	Site	Co	G	L	S	Pe	Da	Ot	C	TM	TK	TP	Cn	Le	Wi	He	Rd	We	Cu	Sn	As	Pb	Mu	Ref
259	Pyr	0	2	0	0	0	0	1	0	0	0	0	1	0	0	0	0	0	0	0	0	0	2585	Evely, 93, 10, no 126
243	Rou	0	3	2	9	0	0	1	1	0	0	0	1	0	0	0	0	0	0	0	0	0	2483	Evely, 93, 6, no 42
248	Rou	0	3	2	9	0	0	1	1	0	0	0	1	0	0	0	0	0	0	0	0	0	2479	Evely, 93, 8, no 62
249	Rou	0	3	2	9	0	0	1	1	0	0	0	1	0	0	0	0	0	0	0	0	0	2480	Evely, 93, 8, no 63
247	Rou	0	3	2	9	0	0	1	1	0	0	0	1	0	0	0	0	0	0	0	0	0	2478	Evely, 93, 8, no 61
250	Rou	0	3	2	9	0	0	1	1	0	0	0	1	0	0	0	0	0	0	0	0	0	2473	Evely, 93, 10, no 123
242	Rou	0	3	2	9	0	0	1	1	0	0	0	1	0	0	0	0	0	0	0	0	0	2482	Evely, 93, 6, no 41
244	Rou	0	3	2	9	0	0	1	1	0	0	0	1	0	0	0	0	0	0	0	0	0	2475	Evely, 93, 8, no 58
246	Rou	0	3	2	9	0	0	1	1	0	0	0	1	0	0	0	0	0	0	0	0	0	2477	Evely, 93, 8, no 60
245	Rou	0	3	2	9	0	0	1	1	0	0	0	1	0	0	0	0	0	0	0	0	0	2476	Evely, 93, 8, no 59
241	Rou	0	3	2	9	0	0	1	1	0	0	0	1	0	0	0	0	0	0	0	0	0	2481	Evely, 93, 6, no 40
1571	Sel	0	1	2	0	0	0	1	1	0	0	0	1	26	0	0	0	300	96.1	0.8	1	0	AshM	Evely, 93,10,no100; Northover / Evely.95, 87, no2, 95
1572	Sel	0	1	2	0	0	0	1	1	0	0	0	1	28	0	0	0	350	0	0	0	0	AshM	Evely, 93, 10, no 101
1570	Sel	0	1	2	0	0	0	1	1	0	0	0	1	25	0	0	0	275	91.6	2.8	2.1	0.1	AshM	Evely,93,10,no102; Northover / Evely 95, 85, no1, 95
256	Sph	0	1	2	6	0	0	1	1	0	0	0	1	0	0	0	0	0	0	0	0	0	2719	Evely, 93, 10, no 121
257	Sph	0	1	2	1	0	0	1	1	0	0	0	1	0	0	0	0	0	0	0	0	0	2720	Evely, 93, 10, no 122
251	Syb	0	4	2	5	0	0	1	1	0	0	0	1	12	0.7	0	0	80	93	7	0	0	AshM	Evely, 93, 8, no 52
232	Tou	0	1	2	4	0	0	1	1	0	0	0	2	32	0	0	0	0	0	0	0	0	539	Evely, 93, 10, no 105
233	Tou	0	1	2	4	0	0	1	1	0	0	0	1	0	0	0	0	0	0	0	0	0	540	Evely, 93, 10, no 106
223	Tyl	0	4	2	4	0	0	1	1	0	0	0	1	17	1.4	0	0	0	0	0	0	0	0	Evely, 93, 8, no 74
224	Tyl	0	4	2	4	0	0	1	1	0	0	0	1	19	1	0	0	0	0	0	0	0	0	Evely, 93, 10, no 91
225	Tyl	0	4	2	4	0	0	1	1	0	0	0	1	11	0.4	0	0	0	0	0	0	0	0	Evely, 93, 10, no 135
261	Vav	0	0	0	0	0	0	1	1	0	0	0	1	0	0	0	0	0	0	0	0	0	2457	Evely, 93, 10, no 128
262	Vav	0	0	0	0	0	0	1	1	0	0	0	1	0	0	0	0	0	0	0	0	0	2458	Evely, 93, 10, no 129
263	Vav	0	0	0	0	0	0	1	1	0	0	0	1	0	0	0	0	0	0	0	0	0	2459	Evely, 93, 10, no 130
260	Vav	0	0	0	0	0	0	1	1	0	0	0	1	0	0	0	0	0	0	0	0	0	2456	Evely, 93, 10, no 127
143	Gou	0	1	1	3	1	0	1	1	0	0	0	1	0	0	0	0	0	0	0	0	0	928	Evely, 93, 10, no 117
141	Gou	F 19	1	1	3	1	0	1	1	0	0	0	1	27	1.2	0	0	0	0	0	0	0	560	Evely, 93, 8, no 86
137	Gou	0	1	1	3	1	0	1	1	0	0	0	1	20	1.2	0	0	0	0	0	0	0	MeM	Evely, 93, 8, no 50
145	Gou	0	1	1	3	1	0	1	1	0	0	0	1	0	0	0	0	0	0	0	0	0	1149	Evely, 93, 10, no 119
136	Gou	0	1	1	3	1	0	1	1	0	0	0	1	0	0	0	0	0	0	0	0	0	967	Evely, 93, 6, no 33
144	Gou	0	1	1	3	1	0	1	1	0	0	0	1	0	0	0	0	0	0	0	0	0	929	Evely, 93, 10, no 118
146	Gou	Cg	1	1	3	1	0	1	1	0	0	0	1	0	0	0	0	0	0	0	0	0	0	Evely, 93, 11, no 147
132	Gou	Hill House	1	1	3	1	0	1	1	0	0	0	1	7.5	0.4	0	0	0	0	0	0	0	0	Evely, 93, 2, no 3
133	Gou	F 18	1	1	3	1	0	1	1	0	0	0	1	20	1.4	0	0	0	0	0	0	0	562	Evely, 93, 6, no 26
142	Gou	0	1	1	3	1	0	1	1	0	0	0	1	0	0	0	0	0	0	0	0	0	561	Evely, 93, 10, no 116
140	Gou	F 18	1	1	3	1	0	1	1	0	0	0	1	27	1.3	0	0	0	0	0	0	0	561	Evely, 93, 8, no 85
138	Gou	0	1	1	3	1	0	1	1	0	0	0	1	0	0	0	0	0	0	0	0	0	563	Evely, 93, 8, no 56
139	Gou	F 18	1	1	3	1	0	1	1	0	0	0	1	32	1	0	0	0	0	0	0	0	559	Evely, 93, 8, no 84
135	Gou	0	1	1	3	1	0	1	1	0	0	0	1	0	0	0	0	0	0	0	0	0	566	Evely, 93, 6, no 32

Appendix V.2 Bronze objects. Sorted by object type / period / site. Total 1857

No	Site	Co	G	L	S	Pe	Da	Ot	C	TM	TK	TP	Cn	Le	Wi	He	Rd	We	Cu	Sn	As	Pb	Mu	Ref
134	Gou	F 18	1	1	3	1	0	1	1	0	0	0	1	16	1	0	0	0	0	0	0	0	563	Evely, 93, 6, no 27
165	Kno	Roy. Road	2	2	2	1	4	1	1	0	0	0	1	3.3	0.2	0	0	0	0	0	0	0	0	Evely, 93, 6, no 8
172	Kno	NW House	2	2	2	1	0	1	1	0	0	0	1	35	1.3	0	0	0	0	0	0	0	0	Evely, 93, 8, no 82
164	Kno	Roy. Road	2	2	2	1	4	1	1	0	0	0	1	3.1	0.2	0	0	0	0	0	0	0	0	Evely, 93, 6, no 7
180	Kno	Roy. Road	2	2	2	1	4	1	1	0	0	0	3	2.7	0.8	0	0	0	0	0	0	0	0	Evely, 93, 11, no 169
177	Kno	Roy. Road	2	2	2	1	4	1	1	0	0	0	1	1.7	0.9	0	0	0	0	0	0	0	0	Evely, 93, 11, no 166
196	Kom	0	3	1	3	1	0	1	1	0	0	0	1	8.9	0.7	0	0	0	0	0	0	0	0	Blitzer, 95, 513, no M97
183	Mal	Quart. D	2	1	2	1	0	1	1	0	0	0	1	0	0	0	0	0	0	0	0	0	0	Evely, 93, 6, no 19
191	Mal	0	2	1	1	1	0	1	1	0	0	0	2	8	0.4	0	0	0	0	0	0	0	0	Evely, 93, 11, no 139
187	Mal	Quart. Z	2	1	2	1	0	1	1	0	0	0	1	25	1.5	0	0	0	0	0	0	0	2934	Evely, 93, 8, no 83
184	Mal	Quart. Z	2	1	2	1	0	1	1	0	0	0	1	3	0.4	0	0	0	0	0	0	0	0	Evely, 93, 6, no 43
186	Mal	Quart. Z	2	1	2	1	0	1	1	0	0	0	2	5.6	0.9	0	0	0	0	0	0	0	0	Evely, 93, 6, no 45
192	Mal	Quart. Z	2	1	2	1	3	1	1	0	0	0	1	2.3	0.2	0	0	0	0	0	0	0	0	Evely, 93, 5, no 2
189	Mal	XXV 2	2	1	1	1	0	1	1	0	0	0	1	21	1	0	0	0	0	0	0	0	2101	Evely, 93, 10, no 88
185	Mal	XXV 2	2	1	1	1	0	1	1	0	0	0	1	4.8	0.6	0	0	0	0	0	0	0	2102	Evely, 93, 6, no 44
182	Mal	Quart. D	2	1	2	1	0	1	1	0	0	0	1	10	0.5	0	0	0	0	0	0	0	2261	Evely, 93, 6, no 18
190	Mal	West Qu	2	1	1	1	0	1	1	0	0	0	1	8.5	0.7	0	0	0	0	0	0	0	0	Evely, 93, 10, no 138
188	Mal	XXV 2	2	1	1	1	0	1	1	0	0	0	1	27	0.7	0	0	0	0	0	0	0	2100	Evely, 93, 8, no 87
1656	Moc	House C3	1	1	3	1	4	1	1	0	0	0	1	0	0	0	0	0	0	0	0	0	0	Soles / Davaras, 96, 196, no CA 54
1735	Moc	House B1	1	1	3	1	4	1	1	0	0	0	1	4.1	0	0	0	0	0	0	0	0	0	Soles / Davaras, 94, 425, no CA 17
1747	Moc	House C3	1	1	3	1	4	1	1	0	0	0	1	0	0	0	0	0	0	0	0	0	0	Soles / Davaras, 96, 194, no CA 52
1748	Moc	House C3	1	1	3	1	4	1	1	0	0	0	1	0	0	0	0	0	0	0	0	0	0	Soles / Davaras, 96, 196, no CA 53
1817	Pak	Block D/5	1	1	3	1	4	1	1	0	0	0	1	0	0	0	0	0	0	0	0	0	0	Driessen / Macdonald, 97, 233
157	Pak	Block Chi	1	1	3	1	4	1	1	0	0	0	1	0	0	0	0	0	0	0	0	0	0	Evely, 93, 6, no 25
154	Pak	Block Chi	1	1	3	1	4	1	1	0	0	0	1	6.5	0.3	0	0	0	0	0	0	0	0	Evely, 93, 6, no 5
163	Pak	Town	1	1	3	1	0	1	1	0	0	0	2	12	1	0	0	0	0	0	0	0	0	Evely, 93, 11, no 164
1363	Pak	Block Chi	1	1	3	1	4	1	1	0	0	0	1	0	0	0	0	0	0	0	0	0	0	Dawkins, 04-05, 282
1364	Pak	Block Chi	1	1	3	1	4	1	1	0	0	0	1	0	0	0	0	0	0	0	0	0	0	Dawkins, 04-05, 282
156	Pak	Block Chi	1	1	3	1	4	1	1	0	0	0	1	16	0	0	0	0	0	0	0	0	0	Evely, 93, 6, no 24
161	Pak	House B	1	1	3	1	0	1	1	0	0	0	1	0	0	0	0	0	0	0	0	0	0	Evely, 93, 11, no 151
162	Pak	House C	1	1	3	1	4	1	1	0	0	0	1	26	3	0	0	0	0	0	0	0	0	Evely, 93, 11, no 163
159	Pak	House C	1	1	3	1	4	1	1	0	0	0	1	34	0	0	0	0	0	0	0	0	0	Evely, 93, 10, no 90
155	Pak	Block Chi	1	1	3	1	4	1	1	0	0	0	1	0	0	0	0	0	0	0	0	0	0	Evely, 93, 6, no 6
158	Pak	Block Ksi	1	1	3	1	4	1	1	0	0	0	1	24	1.3	0	0	0	0	0	0	0	0	Evely, 93, 10, no 89
231	PaZ	0	1	2	4	1	0	1	1	0	0	0	1	17	0	0	0	0	0	0	0	0	2905	Evely, 93, 11, no 162
153	Pha	0	1	3	2	1	4	1	1	0	0	0	1	0	0	0	0	0	0	0	0	0	0	Evely, 93, 11, no 150
148	Pha	0	1	3	2	1	4	1	1	0	0	0	1	13	0	0	0	0	0	0	0	0	0	Evely, 93, 6, no 16
147	Pha	0	1	3	2	1	0	1	1	0	0	0	1	6	0	0	0	0	0	0	0	0	0	Evely, 93, 6, no 4
149	Pha	0	1	3	2	1	0	1	1	0	0	0	2	14	0	0	0	0	0	0	0	0	0	Evely, 93, 6, no 17
152	Pha	0	1	3	2	1	0	1	1	0	0	0	1	0	0	0	0	0	0	0	0	0	0	Evely, 93, 11, no 149

Appendix V.2 / 4

Appendix V.2 Bronze objects. Sorted by object type / period / site. Total 1857

No	Site	Co	G	L	S	Pe	Da	Ot	C	TM	TK	TP	Cn	Le	Wi	He	Rd	We	Cu	Sn	As	Pb	Mu	Ref
151	Pha	0	3	2	1	1	0	1	1	0	0	0	2	14	0	0	0	0	0	0	0	0	0	Evely, 93, 11, no 148
150	Pha	0	3	2	1	1	0	1	1	0	0	0	1	13	0	0	0	0	0	0	0	0	0	Evely, 93, 6, no 46
218	Pse	0	1	1	3	1	0	1	1	0	0	0	1	0	0	0	0	0	0	0	0	0	0	Evely, 93, 6, no 39
217	Pse	0	1	1	3	1	0	1	1	0	0	0	1	0	0	0	0	0	0	0	0	0	0	Evely, 93, 6, no 38
219	Pse	0	1	1	3	1	0	1	1	0	0	0	1	21	2.3	0	0	0	0	0	0	0	0	Evely, 93, 8, no 51
213	Pse	0	1	1	3	1	0	1	1	0	0	0	1	0	0	0	0	0	0	0	0	0	1593	Evely, 93, 6, no 34
220	Pse	0	1	1	3	1	0	1	1	0	0	0	1	22	1.2	0	0	0	0	0	0	0	1594	Evely, 93, 10, no 136
215	Pse	0	1	1	3	1	0	1	1	0	0	0	1	0	0	0	0	0	0	0	0	0	0	Evely, 93, 6, no 36
214	Pse	0	1	1	3	1	0	1	1	0	0	0	1	0	0	0	0	0	0	0	0	0	0	Evely, 93, 6, no 35
216	Pse	0	1	1	3	1	0	1	1	0	0	0	1	0	0	0	0	0	0	0	0	0	1598	Evely, 93, 6, no 37
12	Zak	XIII	1	1	1	1	4	1	1	0	0	0	2	21	4.3	0	0	120	0	0	0	0	0	Platon, 88, II, 194, no 3
14	Zak	XXV	1	1	1	1	4	1	1	0	0	0	1	16	3.4	0	0	40	0	0	0	0	0	Platon, 88, II, 194, no 5
10	Zak	XLIV	1	1	1	1	4	1	1	0	0	0	1	23	3.2	0	0	210	0	0	0	0	2907	Platon, 88, II, 193, no 1
81	Zak	XXVIII	1	1	1	1	4	1	1	0	0	0	1	9.6	0.9	0	0	4	0	0	0	0	2911	Platon, 88, II, 215, no 2
19	Zak	XXVIII	1	1	1	1	4	1	1	0	0	0	2	19	1.3	0	0	180	0	0	0	0	2919	Platon, 88, II, 196, no 10
11	Zak	XIII	1	1	1	1	4	1	1	0	0	0	2	11	4.5	0	0	35	0	0	0	0	0	Platon, 88, II, 194, no 2
80	Zak	XXVIII	1	1	1	1	4	1	1	0	0	0	1	8.6	0.9	0	0	0	0	0	0	0	0	Platon, 88, II, 214, no 1
17	Zak	XXVIII	1	1	1	1	4	1	1	0	0	0	2	15	2.8	0	0	160	0	0	0	0	0	Platon, 88, II, 195, no 8
20	Zak	House B	1	1	2	1	4	1	1	0	0	0	1	19	1.3	0	0	150	0	0	0	0	0	Platon, 88, II, 196, no 11
16	Zak	XXVIII	1	1	1	1	4	1	1	0	0	0	1	26	3.6	0	0	320	0	0	0	0	0	Platon, 88, II, 195, no 7
15	Zak	XXV	1	1	1	1	4	1	1	0	0	0	2	10	3.5	0	0	50	0	0	0	0	0	Platon, 88, II, 195, no 6
18	Zak	XXVIII	1	1	1	1	4	1	1	0	0	0	3	9	2.3	0	0	110	0	0	0	0	0	Platon, 88, II, 195, no 9
13	Zak	XIII	1	1	1	1	4	1	1	0	0	0	1	21	3.2	0	0	53	0	0	0	0	0	Platon, 88, II, 194, no 4
227	Zir	0	1	2	5	2	0	1	1	0	0	0	1	0	0	0	0	0	0	0	0	0	2444	Evely, 93, 10, no 112
226	Zir	0	1	2	5	2	0	1	1	0	0	0	1	23	1	0	0	0	0	0	0	0	2443	Evely, 93, 10, no 111
1857	Kno	UM L	2	2	2	2	5	1	1	0	0	0	2	13	0	0	0	0	92	8	0	0	0	Popham, 84, 51, no L 129
166	Kno	UM M	2	2	2	2	5	1	1	0	0	0	1	11	0.5	0	0	0	90	10	0	0	0	Evely, 93, 6, no 20
168	Kno	UM M	2	2	2	2	5	1	1	0	0	0	1	21	2	0	0	0	91	9	0	0	0	Evely, 93, 6, no 29
170	Kno	N end	2	2	2	2	5	1	1	0	0	0	1	6.8	0.9	0	0	0	0	0	0	0	0	Evely, 93, 8, no 70
1853	Kno	UM M	2	2	2	2	5	1	1	0	0	0	1	43	4.8	0	0	0	90	10	0	0	0	Popham, 84, 57, no M 9
1855	Kno	UM M	2	2	2	2	5	1	1	0	0	0	2	21	0	0	0	0	91	9	0	0	0	Popham, 84, 59, no M 29
1856	Kno	UM M	2	2	2	2	5	1	1	0	0	0	2	11	0	0	0	0	90	10	0	0	0	Popham, 84, 59, no M 30
175	Kno	UM L	2	2	2	2	5	1	1	0	0	0	1	13	0.5	0	0	0	92	8	0	0	0	Evely, 93, 10, no 137
1854	Kno	UM M	2	2	2	2	5	1	1	0	0	0	1	22	0	0	0	0	0	0	0	0	0	Popham, 84, 59, no M 26
173	Kno	UM M	2	2	2	2	5	1	1	0	0	0	1	42	1.5	0	0	0	90	10	0	0	0	Evely, 93, 10, no 94
167	Kno	UM M	2	2	2	2	5	1	1	0	0	0	1	9	0.7	0	0	25	91	9	0	0	0	Evely, 93, 6, no 21
1718	Ath	Lakkos	2	2	4	3	0	1	1	0	0	0	1	0	0	0	0	0	0	0	0	0	HerM	Sakellarakis / Sakellarakis, 97, 602
1717	Ath	Tourkoyet	2	2	4	3	0	1	1	0	0	0	1	0	0	0	0	0	0	0	0	0	HerM	Sakellarakis / Sakellarakis, 97, 602
1707	Cha	0	4	1	3	3	7	1	1	0	0	0	3	0	0	0	0	0	89	10.2	0.2	0	ChaM	Stos-Gale et.al., 00,213,Tabl.2,214,Tabl 3,no 16
1420	EkP	Khritos L	2	2	6	3	0	1	1	0	0	0	1	0	0	0	0	0	0	0	0	0	0	Kanta, 80, 66

Appendix V.2 Bronze objects. Sorted by object type / period / site. Total 1857

No	Site	Co	G	L	S	Pe	Da	Ot	C	TM	TK	TP	Cn	Le	Wi	He	Rd	We	Cu	Sn	As	Pb	Mu	Ref
1415	EkP	Hag. Apos.	2	2	6	3	0	1	1	0	0	0	1	19	0	0	0	0	0	0	0	0	0	Kanta, 80, 59
200	Kal	Tomb 2	4	1	6	3	7	1	1	0	0	0	1	19	0	0	0	0	0	0	0	0	0	Evely, 93, 10, no 96
197	Kal	Tomb 4	4	1	6	3	8	1	1	0	0	0	1	6.3	0.3	0	0	0	0	0	0	0	0	Evely, 93, 6, no 10
202	Kal	Tomb 2	4	1	6	3	7	1	1	0	0	0	1	17	0.9	0	0	0	0	0	0	0	0	Evely, 93, 11, no 143
198	Kal	Tomb 4	4	1	6	3	8	1	1	0	0	0	1	12	0.5	0	0	0	0	0	0	0	0	Evely, 93, 6, no 22
201	Kal	Tomb 4	4	1	6	3	8	1	1	0	0	0	1	25	1.2	0	0	0	0	0	0	0	0	Evely, 93, 10, no 114
199	Kal	Tomb 4	4	1	6	3	8	1	1	0	0	0	1	11	0.4	0	0	0	0	0	0	0	0	Evely, 93, 6, no 23
193	Kom	Hill Top	3	1	3	3	8	1	1	0	0	0	1	16	0.9	0	0	0	0	0	0	0	0	Blitzer, 95, 511, no M61
194	Kom	Hill Top	3	1	3	3	8	1	1	0	0	0	1	12	1.1	0	0	55	0	0	0	0	0	Blitzer, 95, 516, no M147
195	Kom	Cent Hill	3	1	3	3	7	1	1	0	0	0	1	0	0	0	0	0	0	0	0	0	0	Blitzer, 95, 516, no M150
1430	Lig	Tomb 1	3	2	6	3	0	1	1	0	0	0	1	0	0	0	0	0	0	0	0	0	0	Kanta, 80, 84
1425	Tef	0	2	2	6	3	8	1	1	0	0	0	1	0	0	0	0	0	0	0	0	0	0	Kanta, 80, 80
204	ZaP	Tomb 33	2	2	6	3	0	1	1	0	0	0	1	24	1	0	0	55	0	0	0	0	0	Evely, 93, 10, no 95
203	ZaP	Tomb 76	2	2	6	3	7	1	1	0	0	0	1	7.4	0	0	0	0	0	0	0	0	0	1523	Evely, 93, 6, no 9
1292	0	0	0	0	0	0	0	2	1	0	0	0	2	29	0	0	0	0	0	0	0	0	AshM	Catling, 68, 95, no 10
1291	0	0	0	0	0	0	0	2	1	0	0	0	2	22	0	0	0	0	0	0	0	0	AshM	Catling, 68, 94, no 9
1579	Kom	0	3	1	3	0	0	2	1	0	0	0	2	15	1.7	0	0	0	0	0	0	0	0	Blitzer, 95, 515, no M 128
1580	Kom	0	3	1	3	0	0	2	1	0	0	0	2	20	1.2	0	0	0	0	0	0	0	0	Blitzer, 95, 516, no M 144
1578	Kom	0	3	1	3	0	0	2	1	0	0	0	3	5	1.6	0	0	0	0	0	0	0	0	Blitzer, 95, 514, no M 124
1581	Kom	N House	3	1	3	0	0	2	1	0	0	0	1	22	2.3	0	0	0	0	0	0	0	0	Blitzer, 95, 517, no M 161
1327	Pha	0	3	2	1	0	0	2	1	0	0	0	2	0	0	0	0	0	0	0	0	0	0	Sandars, 55, 192
1326	Pha	0	3	2	1	0	0	2	1	0	0	0	1	0	0	0	0	0	0	0	0	0	0	Sandars, 55, 192
1328	Pha	0	3	2	1	0	0	2	1	0	0	0	1	0	0	0	0	0	0	0	0	0	0	Sandars, 55, 192
1319	Psy	0	1	2	7	0	0	2	1	0	0	0	1	14	0	0	0	0	0	0	0	0	444	Boardman, 61, 22, no 61
1317	Psy	0	1	2	7	0	0	2	1	0	0	0	2	13	0	0	0	0	0	0	0	0	445	Boardman, 61, 24
1310	Psy	0	1	2	7	0	0	2	1	0	0	0	1	14	0	0	0	0	0	0	0	0	325	Boardman, 61, 22
1308	Psy	0	1	2	7	0	0	2	1	0	0	0	1	21	0	0	0	0	0	0	0	0	441	Boardman, 61, 22
1320	Psy	0	1	2	7	0	0	2	1	0	0	0	1	11	0	0	0	0	0	0	0	0	446	Boardman, 61, 22, no 61
1316	Psy	0	1	2	7	0	0	2	1	0	0	0	2	8	0	0	0	0	0	0	0	0	323	Boardman, 61, 24
1313	Psy	0	1	2	7	0	0	2	1	0	0	0	1	20	0	0	0	0	0	0	0	0	442	Boardman, 61, 24
1312	Psy	0	1	2	7	0	0	2	1	0	0	0	1	19	0	0	0	0	0	0	0	0	HerM	Boardman, 61, 23
1318	Psy	0	1	2	7	0	0	2	1	0	0	0	1	18	0	0	0	0	0	0	0	0	443	Boardman, 61, 22, no 61
1311	Psy	0	1	2	7	0	0	2	1	0	0	0	1	21	0	0	0	0	0	0	0	0	HerM	Boardman, 61, 22
1307	Psy	0	1	2	7	0	0	2	1	0	0	0	1	13	0	0	0	0	0	0	0	0	AshM	Boardman, 61, 19, no 71
1309	Psy	0	1	2	7	0	0	2	1	0	0	0	1	19	0	0	0	0	0	0	0	0	HerM	Boardman, 61, 22
1315	Psy	0	1	2	7	0	0	2	1	0	0	0	1	13	0	0	0	0	0	0	0	0	1065	Boardman, 61, 24
1314	Psy	0	1	2	7	0	0	2	1	0	0	0	1	11	0	0	0	0	0	0	0	0	322	Boardman, 61, 24
1603	Gou	0	1	1	3	1	4	2	1	0	0	0	1	8	0	0	0	0	0	0	0	0	935	Boyd Hawes, 08, 34, no 20
1614	Gou	0	1	1	3	1	4	2	1	0	0	0	1	13	0	0	0	0	0	0	0	0	604	Boyd Hawes, 08, 34, no 62
1604	Gou	Ei	1	1	3	1	4	2	1	0	0	0	1	9	0	0	0	0	0	0	0	0	940	Boyd Hawes, 08, 34, no 21

Appendix V.2 Bronze objects. Sorted by object type / period / site. Total 1857

No	Site	Co	G	L	S	Pe	Da	Ot	C	TM	TK	TP	Cn	Le	Wi	He	Rd	We	Cu	Sn	As	Pb	Mu	Ref
1602	Gou	0	1	1	3	1	4	2	1	0	0	0	1	17	0	0	0	0	0	0	0	0	601	Boyd Hawes, 08, 34, no 19
1170	Iso	Iso Dep	2	2	6	1	0	2	1	0	0	0	3	0	0	0	0	0	0	0	0	0	0	Evans, 14, 4
1266	Kep	0	2	2	6	1	6	2	1	0	0	0	1	0	0	0	0	0	0	0	0	0	0	Hutchinson, 56, 78,79, no 6
1492	Kno	S House	2	2	2	1	3	2	1	0	0	0	1	23	0	0	0	0	0	0	0	0	0	Georgiou, 79, 22, no 9
1493	Kno	S House	2	2	2	1	3	2	1	0	0	0	1	12	0	0	0	0	0	0	0	0	0	Georgiou, 79, 22, no 10
1574	Kom	0	3	1	3	1	1	2	1	0	0	0	2	10	1.7	0	0	0	0	0	0	0	0	Blitzer, 95, 511, no M 63
1575	Kom	0	3	1	3	1	0	2	1	0	0	0	3	2.1	2	0	0	0	0	0	0	0	0	Blitzer, 95, 511, no M 65
1494	Mal	Quart D	2	1	2	1	0	2	1	0	0	0	2	25	0	0	0	0	0	0	0	0	2254	Georgiou, 79, 29, no 4
1495	Mal	Quart D	2	1	2	1	0	2	1	0	0	0	3	9	0	0	0	0	0	0	0	0	2256	Georgiou, 79, 29, no 5
1803	Mal	Quat.Z, Zg	2	1	2	1	0	2	1	0	0	0	1	0	0	0	0	0	0	0	0	0	0	Driessen / Macdonald, 97, 190
1760	Moc	Build. B	3	1	3	1	4	2	1	0	0	0	1	0	0	0	0	0	0	0	0	0	0	Soles / Davaras, 96, 205, no CA 105
1749	Moc	House C3	3	1	3	1	4	2	1	0	0	0	1	0	0	0	0	0	0	0	0	0	0	Soles / Davaras, 96, 196, no CA 55
1729	Moc	House C3	3	1	3	1	4	2	1	0	0	0	1	0	0	0	0	0	0	0	0	0	0	Soles / Davaras, 94, 402, no CA 66
1750	Moc	House C3	3	1	3	1	4	2	1	0	0	0	1	0	0	0	0	0	0	0	0	0	0	Soles / Davaras, 96, 196, no CA 59
1754	Moc	House C7	3	1	3	1	4	2	1	0	0	0	3	0	0	0	0	0	0	0	0	0	0	Soles / Davaras, 96, 201, no CA 77
1755	Moc	House C7	3	1	3	1	4	2	1	0	0	0	3	0	0	0	0	0	0	0	0	0	0	Soles / Davaras, 96, 201, no CA 78
1507	NiC	Room 7	2	2	4	1	0	2	1	0	0	0	1	22	0	0	0	0	0	0	0	0	2667	Georgiou, 79, 43, no 5
1377	Pak	Kour.	1	1	3	1	4	2	1	0	0	0	3	6	0	0	0	0	0	0	0	0	0	Dawkins / Tod, 02-03, 333
1383	Pak	House N	1	1	6	1	4	2	1	0	0	0	2	13	0	0	0	0	0	0	0	0	0	Sackett / Popham, 63, 300, no 5
1375	Pak	Block L	1	1	3	1	4	2	1	0	0	0	1	0	0	0	0	0	0	0	0	0	0	Driessen / Macdonald, 97, 230
1321	Pse	0	1	1	3	1	0	2	1	0	0	0	1	0	0	0	0	0	0	0	0	0	1585	Sandars, 55, 189
29	Zak	House D	1	1	2	1	4	2	1	0	0	0	1	12	2.4	0	0	0	0	0	0	0	0	Platon, 88, II, 198, no 9
33	Zak	House E	1	1	2	1	4	2	1	0	0	0	2	11	1.8	0	0	0	0	0	0	0	661	Platon, 88, II, 199, no 13
86	Zak	XLIX	1	1	1	1	4	2	1	0	0	0	1	5.9	0.4	0	0	0	0	0	0	0	0	Platon, 88, II, 216, no 4
31	Zak	C.court	1	1	1	1	4	2	1	0	0	0	3	0	4	0	0	0	0	0	0	0	0	Platon, 88, II, 199, no 11
25	Zak	III	1	1	1	1	4	2	1	0	0	0	1	22	4	0	0	55	0	0	0	0	0	Platon, 88, II, 197, no 5
24	Zak	XI	1	1	1	1	4	2	1	0	0	0	2	19	3.4	0	0	0	0	0	0	0	0	Platon, 88, II, 197, no 4
23	Zak	XXV	1	1	1	1	0	2	1	0	0	0	2	15	2.6	0	0	55	0	0	0	0	0	Platon, 88, II, 197, no 3
84	Zak	House B	1	1	2	1	4	2	1	0	0	0	1	6.5	1	0	0	0	0	0	0	0	0	Platon, 88, II, 216, no 2
32	Zak	House I	1	1	2	1	4	2	1	0	0	0	2	15	2.8	0	0	30	0	0	0	0	655	Platon, 88, II, 199, no 12
30	Zak	House Poly.	1	1	2	1	4	2	1	0	0	0	3	7.3	2.7	0	0	30	0	0	0	0	0	Platon, 88, II, 199, no 10
22	Zak	House B	1	1	2	1	4	2	1	0	0	0	1	8	1.8	0	0	0	0	0	0	0	0	Platon, 88, II, 197, no 2
87	Zak	Buil. NW	1	1	2	1	0	2	1	0	0	0	1	5.4	0.7	0	0	0	0	0	0	0	0	Platon, 88, II, 216, no 5
85	Zak	House B	1	1	2	1	0	2	1	0	0	0	2	5	0.8	0	0	0	0	0	0	0	0	Platon, 88, II, 216, no 3
28	Zak	Harb. Road	1	1	2	1	4	2	1	0	0	0	1	13	2	0	0	15	0	0	0	0	0	Platon, 88, II, 198, no 8
26	Zak	XXVIII	1	1	1	1	4	2	1	0	0	0	2	27	3.6	0	0	50	0	0	0	0	0	Platon, 88, II, 197, no 6
83	Zak	House B	1	1	2	1	0	2	1	0	0	0	1	5.6	0.8	0	0	0	0	0	0	0	0	Platon, 88, II, 215, no 1
21	Zak	0	1	1	0	1	4	2	1	0	0	0	2	12	2	0	0	0	0	0	0	0	0	Platon, 88, II, 196, no 1
27	Zak	House B	1	1	2	1	4	2	1	0	0	0	1	14	2.3	0	0	20	0	0	0	0	0	Platon, 88, II, 198, no 7
1714	Ath	Tholos A	2	2	6	2	6	2	1	0	0	0	1	0	0	0	0	0	0	0	0	0	HerM	Sakellarakis / Sakellarakis, 97, 600

Appendix V.2 Bronze objects. Sorted by object type / period / site. Total 1857

No	Site	Co	G	L	S	Pe	Da	Ot	C	TM	TK	TP	Cn	Le	Wi	He	Rd	We	Cu	Sn	As	Pb	Mu	Ref
1712	Arh	Bur.Buil. 3	2	2	6	2	6	2	1	0	0	0	1	0	0	0	0	0	0	0	0	0	HerM	Sakellarakis / Sakellarakis, 97, 600
1713	Arh	Bur.Buil. 3	2	2	6	2	6	2	1	0	0	0	1	0	0	0	0	0	0	0	0	0	HerM	Sakellarakis / Sakellarakis, 97, 600
1464	Cha	0	4	1	6	2	6	2	1	0	0	0	1	0	0	0	0	0	0	0	0	0	0	Kanta, 80, 226
1463	Cha	0	4	1	6	2	6	2	1	0	0	0	1	0	0	0	0	0	0	0	0	0	0	Kanta, 80, 226
1325	Gol	0	0	0	0	2	5	2	1	0	0	0	1	0	0	0	0	0	0	0	0	0	AshM	Sandars, 55, 195
1244	Gyp	Tomb I	2	2	6	2	6	2	1	0	0	0	1	23	0	0	0	0	0	0	0	0	0	Hood et al, 58-59, 245, no I.7
1242	Gyp	Tomb I	2	2	6	2	6	2	1	0	0	0	1	34	0	0	0	0	0	0	0	0	0	Hood et al, 58-59, 245, no I.5
1174	Iso	Tomb 2	2	2	6	2	5	2	1	0	0	0	3	11	0	0	0	0	0	0	0	0	0	Evans, 14, 58, no 2c
1177	Iso	Tomb 3	2	2	6	2	6	2	1	0	0	0	2	0	0	0	0	0	0	0	0	0	0	Evans, 14, 15, no 3b
1173	Iso	Tomb 2	2	2	6	2	5	2	1	0	0	0	1	28	0	0	0	0	0	0	0	0	0	Evans, 14, 58, no 2d
1178	Iso	Tomb 3	2	2	6	2	6	2	1	0	0	0	2	0	0	0	0	0	0	0	0	0	0	Evans, 14, 15, no 3c
1481	Kno	UM L	2	2	2	2	5	2	1	0	0	0	1	15	0	0	0	0	0	0	0	0	0	Popham, 84, 45, no L15; Catling / Catling, 84, 213
1480	Kno	UM M	2	2	2	2	5	2	1	0	0	0	1	29	0	0	0	0	0	0	0	0	0	Popham, 84, 61, no M80; Catling / Catling, 84, 213
1482	Kno	UM H	2	2	2	2	5	2	1	0	0	0	3	11	0	0	0	0	88	12	0	0	0	Popham, 84, 29, no H93; Catling / Catling, 84, 213
1483	Kno	UM P	2	2	2	2	5	2	1	0	0	0	1	15	0	0	0	0	0	0	0	0	0	Popham, 84, 78, no P2; Catling / Catling, 84, 213
1223	Mas	Tomb VIIA	2	2	6	2	6	2	1	0	0	0	1	0	0	0	0	0	0	0	0	0	0	Forsdyke, 26-27, 262, no VIIA.9
1239	Mas	Tomb XIX	2	2	6	2	6	2	1	0	0	0	1	0	0	0	0	0	0	0	0	0	0	Forsdyke, 26-27, 282, no XIX.2
1226	Mas	Tomb IXE	2	2	6	2	6	2	1	0	0	0	2	0	0	0	0	0	0	0	0	0	2150	Forsdyke, 26-27, 269, no IXE.5; Sandars, 55, 192
1228	Mas	Tom XVIIA	2	2	6	2	6	2	1	0	0	0	1	0	0	0	0	0	0	0	0	0	0	Forsdyke, 26-27, 278, no XVIIA.1
1235	Mas	Tomb XVIII	2	2	6	2	6	2	1	0	0	0	2	0	0	0	0	0	0	0	0	0	0	Forsdyke, 26-27, 282, no XVIII.4
1240	Mas	Tomb XX	2	2	6	2	6	2	1	0	0	0	2	0	0	0	0	0	0	0	0	0	2154	Forsdyke, 26-27, 283, no XX.1; Sandars, 55, 190
1199	Mas	Tomb III	2	2	6	2	6	2	1	0	0	0	1	0	0	0	0	0	0	0	0	0	2159	Forsdyke, 26-27, 252, no III.4; Sandars, 55, 192
1234	Mas	Tomb XVIII	2	2	6	2	6	2	1	0	0	0	2	0	0	0	0	0	0	0	0	0	2152	Forsdyke, 26-27, 282, no XVIII.4; Sandars, 55, 190
1238	Mas	Tomb XIX	2	2	6	2	6	2	1	0	0	0	1	8.5	0	0	0	0	0	0	0	0	2153	Forsdyke, 26-27, 282, no XIX.2; Sandars, 55, 190
1323	Pak	Block L	1	1	3	2	5	2	1	0	0	0	1	0	0	0	0	0	0	0	0	0	601	Sandars, 55, 193
1462	Pig	0	4	2	6	2	6	2	1	0	0	0	2	0	0	0	0	0	0	0	0	0	0	Kanta, 80, 213
1184	Sel	Tomb 4/II	2	2	6	2	6	2	1	0	0	0	2	9	0	0	0	0	0	0	0	0	0	Catling / Catling, 74, 229, no 8
1185	Sel	Tomb 4/I	2	2	6	2	6	2	1	0	0	0	2	8	0	0	0	0	0	0	0	0	0	Catling / Catling, 74, 229, no 9
1324	Sit	0	1	1	0	2	5	2	1	0	0	0	1	0	0	0	0	0	0	0	0	0	1025	Sandars, 55, 195
1153	ZaP	Tomb 75	2	2	6	2	6	2	1	0	0	0	1	27	0	0	0	0	0	0	0	0	1118	Evans, 06, 77, no 75d; Sandars, 55, 192
1120	ZaP	Tomb 15	2	2	6	2	6	2	1	0	0	0	1	0	0	0	0	0	0	0	0	0	0	Evans, 06, 45, no 15a
1134	ZaP	Tomb 51	2	2	6	2	6	2	1	0	0	0	2	23	0	0	0	0	0	0	0	0	0	Evans, 06, 64, no 51d
1157	ZaP	Tomb 80	2	2	6	2	6	2	1	0	0	0	1	32	0	0	0	0	0	0	0	0	1120	Evans, 06, 80, no 80a; Sandars, 55, 189
1140	ZaP	Tomb 64	2	2	6	2	6	2	1	0	0	0	1	21	0	0	0	0	0	0	0	0	AshM	Evans, 06, 69, no 64d; Baboula / Northover, 99, 148
1112	ZaP	Tomb 10	2	2	6	2	6	2	1	0	0	0	1	23	0	0	0	0	0	0	0	0	0	Evans, 06, 31, no 10b
1101	ZaP	Tomb 1	2	2	6	2	6	2	1	0	0	0	1	12	0	0	0	0	0	0	0	0	0	Evans, 06, 22, no 1c
1103	ZaP	Tomb 3	2	2	6	2	6	2	1	0	0	0	1	30	0	0	0	0	0	0	0	0	1116	Evans, 06, 23, no 3a; Sandars, 55, 192
1107	ZaP	Tomb 4	2	2	6	2	6	2	1	0	0	0	1	36	0	0	0	0	0	0	0	0	1115	Evans, 06, 23, no 4b; Sandars, 55, 189
1139	ZaP	Tomb 64	2	2	6	2	6	2	1	0	0	0	1	34	0	0	0	0	0	0	0	0	1114	Evans, 06, 69, no 64c; Sandars, 55, 189
1132	ZaP	Tomb 43	2	2	6	2	6	2	1	0	0	0	2	32	0	0	0	0	0	0	0	0	0	Evans, 06, 62, no 43c

Appendix V.2   Bronze objects. Sorted by object type / period / site. Total 1857   **Appendix V.2 / 8**

No	Site	Co	G	L	S	Pe	Da	Ot	C	TM	TK	TP	Cn	Le	Wi	He	Rd	We	Cu	Sn	As	Pb	Mu	Ref
1161	ZaP	Tomb 92	2	2	6	2	6	2	1	0	0	0	0	0	0	0	0	0	0	0	0	0	0	Evans, 06, 83, no 83b
1110	ZaP	Tomb 7	2	2	6	2	6	2	1	0	0	0	1	19	0	0	0	0	0	0	0	0	0	Evans, 06, 25, no 7a
1154	ZaP	Tomb 76	2	2	6	2	6	2	1	0	0	0	2	21	0	0	0	0	0	0	0	0	0	Evans, 06, 78, no 76b
1160	ZaP	Tomb 92	2	2	6	2	6	2	1	0	0	0	1	0	0	0	0	0	0	0	0	0	0	Evans, 06, 83, no 83a
1115	ZaP	Tomb 13	2	2	6	2	6	2	1	0	0	0	1	0	0	0	0	0	0	0	0	0	1123	Evans, 06, 34, no 13a; Sandars, 55, 189
1133	ZaP	Tomb 51	2	2	6	2	6	2	1	0	0	0	1	23	0	0	0	0	0	0	0	0	0	Evans, 06, 64, no 51c
1138	ZaP	Tomb 62	2	2	6	2	6	2	1	0	0	0	1	22	0	0	0	0	0	0	0	0	0	Evans, 06, 68, no 62d
1137	ZaP	Tomb 55	2	2	6	2	6	2	1	0	0	0	1	0	0	0	0	0	0	0	0	0	0	Evans, 06, 64, no 55b
1114	ZaP	Tomb 12	2	2	6	2	6	2	1	0	0	0	1	12	0	0	0	0	0	0	0	0	0	Evans, 06, 33, no 12b
1116	ZaP	Tomb 13	2	2	6	2	6	2	1	0	0	0	1	0	0	0	0	0	0	0	0	0	0	Evans, 06, 34 ,no 13b
1119	ZaP	Tomb 14	2	2	6	2	6	2	1	0	0	0	2	0	0	0	0	0	0	0	0	0	0	Evans, 06, 43, no 14r
1135	ZaP	Tomb 51	2	2	6	2	6	2	1	0	0	0	1	16	0	0	0	0	0	0	0	0	0	Evans, 06, 64, no 51f
1102	ZaP	Tomb 1	2	2	6	2	6	2	1	0	0	0	1	12	0	0	0	0	0	0	0	0	1442	Evans, 06, 22, no 1d; Sandars, 55, 189
1158	ZaP	Tomb 80	2	2	6	2	6	2	1	0	0	0	1	20	0	0	0	0	0	0	0	0	0	Evans, 06, 80, no 80b
1159	ZaP	Tomb 81	2	2	6	2	6	2	1	0	0	0	1	15	0	0	0	0	0	0	0	0	0	Evans, 06, 80, no 81a
1416	EkP	Malates.L	2	2	6	3	0	2	1	0	0	0	1	0	0	0	0	0	0	0	0	0	0	Kanta, 80, 60
1359	Gal	0	3	2	6	3	7	2	1	0	0	0	1	0	0	0	0	0	0	0	0	0	0	Driessen / Macdonald, 84, 71, no Eii 4
1399	Gor	Tomb 2	2	2	6	3	8	2	1	0	0	0	1	0	0	0	0	0	0	0	0	0	0	Kanta, 80, 48
1252	Gyp	Tomb VI	2	2	6	3	8	2	1	0	0	0	3	8.4	0	0	0	0	0	0	0	0	0	Hood et al, 58-59, 246, no VI.5
1582	Kom	0	3	1	3	3	0	2	1	0	0	0	2	9.2	3.1	0	0	0	0	0	0	0	0	Blitzer, 95, 517, no M 166
1576	Kom	0	3	1	3	3	8	2	1	0	0	0	3	2.1	1.4	0	0	0	0	0	0	0	0	Blitzer, 95, 512, no M 81
1577	Kom	0	3	1	3	3	0	2	1	0	0	0	3	1.3	1.6	0	0	0	0	0	0	0	0	Blitzer, 95, 512, no M 83
1208	Mas	Tomb V	2	2	6	3	8	2	1	0	0	0	2	0	0	0	0	0	0	0	0	0	2151	Forsdyke, 26-27, 257, no V.2; Sandars, 55, 190
1770	Moc	Tomb 15	1	1	6	3	7	2	1	0	0	0	1	0	0	0	0	0	0	0	0	0	0	Soles / Davaras, 96, 222, no CA 113
1769	Moc	Tomb 15	1	1	6	3	0	2	1	0	0	0	1	0	0	0	0	0	0	0	0	0	0	Soles / Davaras, 96, 222, no CA 110
1402	Moh	0	2	2	6	3	8	2	1	0	0	0	1	0	0	0	0	0	0	0	0	0	0	Kanta, 80, 52
1391	Mon	0	4	2	6	3	7	2	1	0	0	0	1	0	0	0	0	0	0	0	0	0	0	Kanta, 80, 15
1431	Mor	0	3	2	6	3	0	2	1	0	0	0	1	0	0	0	0	0	0	0	0	0	0	Kanta, 80, 84
1655	Pak	Beehive T	1	1	6	3	7	2	1	0	0	0	1	22	0	0	0	0	0	0	0	0	0	Bosanquet, 01-02, 304,;Sandars, 63, 148
1322	Pak	Gournia	1	1	3	3	0	2	1	0	0	0	1	0	0	0	0	0	0	0	0	0	601	Sandars, 55, 190
1445	Phk	0	1	2	6	3	7	2	1	0	0	0	1	0	0	0	0	0	0	0	0	0	0	Kanta, 80, 164, 183
1446	Phk	0	1	2	6	3	7	2	1	0	0	0	1	0	0	0	0	0	0	0	0	0	0	Kanta, 80, 164, 183
1301	Psy	0	1	2	7	3	0	2	1	0	0	0	1	19	0	0	0	0	0	0	0	0	AshM	Boardman, 61, 19, no 65
1294	Psy	0	1	2	7	3	0	2	1	0	0	0	1	13	0	0	0	0	0	0	0	0	AshM	Boardman, 61, 19, no 58
1303	Psy	0	1	2	7	3	0	2	1	0	0	0	1	16	0	0	0	0	94.8	4.6	0.3	0.3	AshM	Boardman, 61, 19, no 67, 160
1295	Psy	0	1	2	7	3	0	2	1	0	0	0	1	14	0	0	0	0	0	0	0	0	AshM	Boardman, 61, 19, no 59
1298	Psy	0	1	2	7	3	0	2	1	0	0	0	1	12	0	0	0	0	0	0	0	0	AshM	Boardman, 61, 19, no 62
1300	Psy	0	1	2	7	3	0	2	1	0	0	0	1	8.1	0	0	0	0	0	0	0	0	AshM	Boardman, 61, 19, no 64
1297	Psy	0	1	2	7	3	0	2	1	0	0	0	1	8.5	0	0	0	0	0	0	0	0	AshM	Boardman, 61, 19, no 61
1296	Psy	0	1	2	7	3	0	2	1	0	0	0	1	10	0	0	0	0	0	0	0	0	AshM	Boardman, 61, 19, no 60

Appendix V.2 Bronze objects. Sorted by object type / period / site. Total 1857

No	Site	Co	G	L	S	Pe	Da	Ot	C	TM	TK	TP	Cn	Le	Wi	He	Rd	We	Cu	Sn	As	Pb	Mu	Ref
1306	Psy	0	1	2	7	3	0	2	1	0	0	0	1	22	0	0	0	0	91	8.2	0.4	0.2	AshM	Boardman, 61, 19, no 70, 160
1305	Psy	0	1	2	7	3	0	2	1	0	0	0	1	9.6	0	0	0	0	88	10.9	0.3	0.7	AshM	Boardman, 61, 19, no 69, 160
1299	Psy	0	1	2	7	3	0	2	1	0	0	0	1	24	0	0	0	0	91.2	5.7	0.4	2	AshM	Boardman, 61, 19, no 63, 160
1302	Psy	0	1	2	7	3	0	2	1	0	0	0	1	7.9	0	0	0	0	0	0	0	0	AshM	Boardman, 61, 19, no 66
1304	Psy	0	1	2	7	3	0	2	1	0	0	0	1	19	0	0	0	0	89	8.2	1.3	0	AshM	Boardman, 61, 19, no 68, 160
1424	Tcf	0	2	2	6	3	8	2	1	0	0	0	1	0	0	0	0	0	0	0	0	0	0	Kanta, 80, 80
378	0	0	0	0	0	0	0	3	1	0	0	0	1	23	4.2	0	0	0	0	0	0	0	BriM	Evely, 93, 47, no 107
466	0	0	0	0	0	0	0	3	1	0	0	0	1	0	0	0	0	0	0	0	0	0	1712	Evely, 93, 49, no 197
370	0	0	0	0	0	0	0	3	1	0	0	0	1	20	0	0	0	0	0	0	0	0	1713	Evely, 93, 46, no 78
376	0	0	0	0	0	0	0	3	1	0	0	0	1	0	0	0	0	0	0	0	0	0	GiaC	Evely, 93, 46, no 105
407	0	0	0	0	0	0	0	3	1	0	0	0	1	0	1	0	0	0	99	1	0	0	0	Evely, 93, 49, no 199
369	0	0	0	0	0	0	0	3	1	0	0	0	1	16	4.9	0	0	0	0	0	0	0	Louv	Evely, 93, 46, no 77
379	0	0	1	0	0	0	0	3	1	0	0	0	2	16	6.1	0	0	0	0	0	0	0	IerM	Evely, 93, 47, no 108
374	0	0	0	0	0	0	0	3	1	0	0	0	1	16	4.5	0	0	0	0	0	0	0	GiaC	Evely, 93, 46, no 103
381	0	0	4	0	0	0	0	3	1	0	0	0	1	19	5.3	0	0	0	0	0	0	0	ChaM	Evely, 93, 46, no 99
372	0	0	0	0	0	0	0	3	1	0	0	0	1	19	0	0	0	0	0	0	0	0	BSA	Evely, 93, 46, no 101
377	0	0	0	0	0	0	0	3	1	0	0	0	1	22	4.4	0	0	0	0	0	0	0	BriM	Evely, 93, 46, no 106
373	0	0	0	0	0	0	0	3	1	0	0	0	1	13	5	0	0	0	0	0	0	0	GiaC	Evely, 93, 46, no 102
371	0	0	0	0	0	0	0	3	1	0	0	0	1	20	0	0	0	0	0	0	0	0	1728	Evely, 93, 46, no 79
382	0	0	4	0	0	0	0	3	1	0	0	0	1	14	5.1	0	0	0	0	0	0	0	ChaM	Evely, 93, 46, no 100
467	0	0	0	0	0	0	0	3	1	0	0	0	1	0	0	0	0	0	0	0	0	0	2430	Evely, 93, 49, no 198
375	0	0	0	0	0	0	0	3	1	0	0	0	1	0	0	0	0	0	0	0	0	0	385	Evely, 93, 46, no 104
429	Amd	0	1	2	0	0	0	3	1	0	0	0	1	12	3.4	0	0	0	0	0	0	0	2162	Evely, 93, 47, no 110
445	Amn	0	2	1	4	0	0	3	1	0	0	0	1	0	0	0	0	0	0	0	0	0	0	Evely, 93, 49, no 169
459	Api	0	1	2	6	0	0	3	1	0	0	0	1	0	0	0	0	0	0	0	0	0	2488	Evely, 93, 49, no 190
447	Arv	0	1	1	5	0	0	3	1	0	0	0	1	0	0	0	0	0	0	0	0	0	0	Evely, 93, 49, no 171
428	Ber	0	0	0	0	0	0	3	1	0	0	0	1	16	5.8	0	0	0	0	0	0	0	2285a	Evely, 93, 47, no 109
462	Elk	0	0	0	0	0	0	3	1	0	0	0	1	0	0	0	0	0	0	0	0	0	2738	Evely, 93, 49, no 193
460	Elk	0	0	0	0	0	0	3	1	0	0	0	1	0	0	0	0	0	0	0	0	0	2736	Evely, 93, 49, no 191
461	Elk	0	0	0	0	0	0	3	1	0	0	0	1	0	0	0	0	0	0	0	0	0	2737	Evely, 93, 49, no 192
441	Gon	0	4	2	5	0	0	3	1	0	0	0	1	0	0	0	0	0	0	0	0	0	1790	Evely, 93, 49, no 152
418	Iep	0	1	1	0	0	0	3	1	0	0	0	1	11	5	0	0	0	0	0	0	0	311	Evely, 93, 44, no 65
430	Iep	0	1	1	0	0	0	3	1	0	0	0	1	0	0	0	0	0	0	0	0	0	299	Evely, 93, 47, no 122
431	Kam	0	3	2	7	0	0	3	1	0	0	0	1	0	0	0	0	0	0	0	0	0	0	Evely, 93, 47, no 123
419	Kek	0	1	1	0	0	0	3	1	0	0	0	1	0	0	0	0	0	0	0	0	0	1547	Evely, 93, 46, no 66
463	Kha	0	0	1	2	0	0	3	1	0	0	0	1	0	0	0	0	0	0	0	0	0	2489	Evely, 93, 49, no 194
450	Kho	0	1	2	0	0	0	3	1	0	0	0	3	0	0	0	0	0	0	0	0	0	985	Evely, 93, 49, no 174
336	Kno	0	2	2	0	0	0	3	1	0	0	0	1	19	5	0	0	0	0	0	0	0	GiaC	Evely, 93, 46, no 81
335	Kno	0	2	2	1	0	0	3	1	0	0	0	1	17	4.1	0	8.8	0	91	6.4	1.4	1.5	AshM	Evely, 93, 46, no67; Northover / Evely, 95, 87, no 5
337	Kno	0	2	2	0	0	0	3	1	0	0	0	1	18	4.7	0	0	0	0	0	0	0	GiaC	Evely, 93, 46, no 82

Appendix V.2 Bronze objects. Sorted by object type / period / site. Total 1857

No	Site	Co	G	L	S	Pe	Da	Ot	C	TM	TK	TP	Cn	Le	Wi	He	Rd	We	Cu	Sn	As	Pb	Mu	Ref
396	Kom	Hill Top	3	1	3	0	0	3	1	0	0	0	1	20	5.6	0	0	0	0	0	0	0	0	Evely, 93, 44, no 57
424	Kri	0	1	2	0	0	0	3	1	0	0	0	1	0	0	0	0	0	0	0	0	0	2269	Evely, 93, 46, no 83
421	Las	0	2	2	0	0	0	3	1	0	0	0	1	0	0	0	0	0	0	0	0	0	1460	Evely, 93, 46, no 69
420	Las	0	2	2	0	0	0	3	1	0	0	0	1	0	0	0	0	0	0	0	0	0	1661	Evely, 93, 46, no 68
465	Mal	0	2	1	0	0	0	3	1	0	0	0	1	0	0	0	0	0	0	0	0	0	4347	Evely, 93, 49, no 196
433	Mar	0	1	2	0	0	0	3	1	0	0	0	1	8	4.3	0	0	0	0	0	0	0	GiaC	Evely, 93, 47, no 125
434	Mel	0	2	2	7	0	0	3	1	0	0	0	1	15	4.4	0	0	0	0	0	0	0	2770	Evely, 93, 47, no 126
425	Mes	0	2	2	0	0	0	3	1	0	0	0	1	11	4	0	0	0	0	0	0	0	GiaC	Evely, 93, 46, no 90
426	Mes	0	2	2	0	0	0	3	1	0	0	0	1	22	4	0	0	0	0	0	0	0	GiaC	Evely, 93, 49, no 141
464	Mes	0	3	2	0	0	0	3	1	0	0	0	1	0	0	0	0	0	0	0	0	0	2589	Evely, 93, 49, no 195
440	Moh	0	2	2	0	0	0	3	1	0	0	0	1	16	2.4	0	0	0	0	0	0	0	GiaC	Evely, 93, 49, no 150
423	Mou	0	2	1	0	0	0	3	1	0	0	0	1	20	4.8	0	0	0	0	0	0	0	0	Evely, 93, 46, no 80
388	Mvr	0	3	2	0	0	0	3	1	0	0	0	1	16	4.3	0	0	0	0	0	0	0	GiaC	Evely, 93, 46, no 89
383	Mvr	0	3	2	0	0	0	3	1	0	0	0	1	16	4.5	0	0	0	0	0	0	0	GiaC	Evely, 93, 46, no 84
384	Mvr	0	3	2	0	0	0	3	1	0	0	0	1	17	4.5	0	0	0	0	0	0	0	GiaC	Evely, 93, 46, no 85
387	Mvr	0	3	2	0	0	0	3	1	0	0	0	1	16	4.3	0	0	0	0	0	0	0	GiaC	Evely, 93, 46, no 88
386	Mvr	0	3	2	0	0	0	3	1	0	0	0	1	16	4.3	0	0	0	0	0	0	0	GiaC	Evely, 93, 46, no 87
385	Mvr	0	3	2	0	0	0	3	1	0	0	0	1	15	5	0	0	0	0	0	0	0	GiaC	Evely, 93, 46, no 86
365	Pak	0	1	1	3	0	0	3	1	0	0	0	3	0	0	0	0	0	82.8	10.9	0.5	0.2	1382	Evely, 93, 49, no 145c, Mangou / Ioannou, 98, 95
367	Pak	Kour.	1	1	3	0	0	3	1	0	0	0	1	0	0	0	0	0	0	0	0	0	0	Evely, 93, 49, no 168
363	Pak	0	1	1	3	0	0	3	1	0	0	0	1	0	0	0	0	0	85.9	1.9	0.4	1	1377	Evely, 93, 49, no 145c; Mangou / Ioannou, 98, 95
364	Pak	0	1	1	3	0	0	3	1	0	0	0	3	0	0	0	0	0	82.9	10	0.2	0.1	1380	Evely, 93, 49, no 145d; Mangou / Ioannou, 98, 95
357	Pak	0	1	1	3	0	0	3	1	0	0	0	1	16	3.6	0	0	0	83.5	7.3	0.1	0.2	1383	Evely, 93, 47, no 120; Mangou / Ioannou, 98, 95
435	Phk	0	1	2	6	0	0	3	1	0	0	0	1	13	4.8	0	0	0	0	0	0	0	HNM	Evely, 93, 47, no 137
452	Pho	0	4	2	0	0	0	3	1	0	0	0	1	0	0	0	0	0	0	0	0	0	1460	Evely, 93, 49, no 180
436	Pin	0	1	2	0	0	0	3	1	0	0	0	1	16	4.5	0	0	0	0	0	0	0	HNM	Evely, 93, 47, no138
397	Pra	0	1	2	6	0	0	3	1	0	0	0	1	17	5.7	0	0	0	0	0	0	0	0	Evely, 93, 46, no 70
394	Pse	0	1	1	3	0	0	3	1	0	0	0	1	20	5	0	0	0	0	0	0	0	1589	Evely, 93, 44, no 64
395	Pse	0	1	1	3	0	0	3	1	0	0	0	1	12	5.8	0	0	0	0	0	0	0	1590	Evely, 93, 47, no 121
392	Psy	0	1	2	7	0	0	3	1	0	0	0	3	6.1	5.3	0	0	0	88	11	0	0	1380	Evely, 93, 42, no 25
390	Psy	0	1	2	7	0	0	3	1	0	0	0	1	16	4.7	0	0	956	91	7	1.1	0.4	AshM	Evely, 93, 42, no 23; Northover / Evely, 95, 87, no 6
391	Psy	0	1	2	7	0	0	3	1	0	0	0	1	17	4.5	0	0	0	0	0	0	0	300	Evely, 93, 42, no 24
451	RoE	0	1	2	0	0	0	3	1	0	0	0	1	0	0	0	0	0	0	0	0	0	1023	Evely, 93, 49, no 178
422	Rog	0	0	0	0	0	0	3	1	0	0	0	1	0	0	0	0	0	0	0	0	0	2265	Evely, 93, 46, no 71
405	Rou	0	3	2	9	0	0	3	1	0	0	0	1	0	0	0	0	0	0	0	0	0	2470	Evely, 93, 49, no 177
404	Rou	0	3	2	9	0	0	3	1	0	0	0	1	0	0	0	0	0	0	0	0	0	2469	Evely, 93, 49, no 176
403	Rou	0	3	2	9	0	0	3	1	0	0	0	1	0	0	0	0	0	0	0	0	0	2468	Evely, 93, 49, no 175
458	Ski	0	1	2	0	0	0	3	1	0	0	0	1	0	0	0	0	0	0	0	0	0	0	Evely, 93, 49, no 189
448	Sta	0	1	2	4	0	0	3	1	0	0	0	1	0	0	0	0	0	0	0	0	0	0	Evely, 93, 49, no 172
442	Sty	0	0	0	0	0	0	3	1	0	0	0	1	0	0	0	0	0	0	0	0	0	2162	Evely, 93, 49, no 153

Appendix V.2 Bronze objects. Sorted by object type / period / site. Total 1857

No	Site	Co	G	L	S	Pe	Da	Ot	C	TM	TK	TP	Cn	Le	Wi	He	Rd	We	Cu	Sn	As	Pb	Mu	Ref
406	Syb	0	4	2	5	0	0	3	1	0	0	0	1	0	0	0	0	0	0	0	0	0	0	Evely, 93, 49, no 179
427	Syk	0	1	2	0	0	0	3	1	0	0	0	1	0	0	0	0	0	0	0	0	0	390	Evely, 93, 46, no 98
449	Sym	0	1	2	8	0	0	3	1	0	0	0	1	0	0	0	0	0	0	0	0	0	0	Evely, 93, 49, no 173
400	Tou	0	1	2	5	0	0	3	1	0	0	0	1	0	0	0	0	0	0	0	0	0	533	Evely, 93, 46, no 74
399	Tou	0	1	2	5	0	0	3	1	0	0	0	1	22	6.6	0	0	0	0	0	0	0	536	Evely, 93, 46, no 73
402	Tou	0	1	2	5	0	0	3	1	0	0	0	1	0	0	0	0	0	0	0	0	0	537	Evely, 93, 46, no 76
401	Tou	0	1	2	5	0	0	3	1	0	0	0	1	0	0	0	0	0	0	0	0	0	534	Evely, 93, 46, no 75
398	Tou	0	1	2	5	0	0	3	1	0	0	0	1	22	5.8	0	0	0	0	0	0	0	535	Evely, 93, 46, no 72
432	Tro	0	1	2	0	0	0	3	1	0	0	0	1	0	0	0	0	0	99	0	0	0	0	Evely, 93, 47, no 124
1851	Vav	0	0	2	0	0	0	3	1	0	0	0	1	0	0	0	0	0	0	0	0	0	2451	Evely, 93, 49, no 183
454	Vav	0	0	2	0	0	0	3	1	0	0	0	1	0	0	0	0	0	0	0	0	0	2450	Evely, 93, 49, no 182
453	Vav	0	0	2	0	0	0	3	1	0	0	0	1	0	0	0	0	0	0	0	0	0	2449	Evely, 93, 49, no 181
457	Vav	0	0	2	0	0	0	3	1	0	0	0	1	0	0	0	0	0	0	0	0	0	2454	Evely, 93, 49, no 186
456	Vav	0	0	2	0	0	0	3	1	0	0	0	1	0	0	0	0	0	0	0	0	0	2453	Evely, 93, 49, no 185
455	Vav	0	0	2	0	0	0	3	1	0	0	0	1	0	0	0	0	0	0	0	0	0	2452	Evely, 93, 49, no 184
380	0	0	1	0	0	1	0	3	1	0	0	0	1	26	0	0	0	0	0	0	0	0	0	Evely, 93, 49, no 157
368	0	0	0	0	0	1	0	3	1	0	0	0	1	17	4.7	0	0	925	95	2.3	2	0.2	AshM	Evely, 93, 44, no 56
389	Apd	0	4	2	5	1	0	3	1	0	0	0	1	12	4	0	0	0	0	0	0	0	2396	Evely, 93, 42, no 8
326	Gou	G 1	1	2	3	1	0	3	1	0	0	0	1	13	5	0	0	0	0	0	0	0	554	Evely, 93, 42, no 12
327	Gou	F 18	1	2	3	1	0	3	1	0	0	0	1	12	4.8	0	0	0	0	0	0	0	555	Evely, 93, 42, no 13
324	Gou	G 1	1	2	3	1	0	3	1	0	0	0	3	7.5	5.1	0	0	0	0	0	0	0	MeM	Evely, 93, 42, no 10
325	Gou	A 23	1	2	3	1	0	3	1	0	0	0	1	18	5	0	0	0	0	0	0	0	553	Evely, 93, 42, no 11
323	Gou	A 23	1	2	3	1	0	3	1	0	0	0	1	18	4.1	0	0	0	0	0	0	0	0	Evely, 93, 42, no 9
1774	HaT	Room 16	3	2	4	1	0	3	1	0	0	0	1	17	0	0	0	0	0	0	0	0	1226	Evely, 93, 44, no 60; Driessen / Macdonald, 97, 202
1775	HaT	Room 16	3	2	4	1	0	3	1	0	0	0	1	17	0	0	0	0	0	0	0	0	1227	Evely, 93, 44, no 61; Driessen / Macdonald, 97, 202
1777	HaT	Room 27	3	2	4	1	0	3	1	0	0	0	1	17	0	0	0	0	0	0	0	0	1532	Evely, 93, 44, no 63; Driessen / Macdonald, 97, 202
1776	HaT	Room 27	3	2	4	1	0	3	1	0	0	0	1	17	0	0	0	0	0	0	0	0	1228	Evely, 93, 44, no 62; Driessen / Macdonald, 97, 202
1773	HaT	Room 16	3	2	4	1	0	3	1	0	0	0	1	17	5.3	0	0	0	0	0	0	0	1225	Evely, 93, 44, no 59; Driessen / Macdonald, 97, 202
1772	HaT	Room 16	3	2	4	1	0	3	1	0	0	0	1	17	5.3	0	0	0	0	0	0	0	1224	Evely, 93, 44, no 58; Driessen / Macdonald, 97, 202
411	Hel	0	4	2	5	1	0	3	1	0	0	0	1	19	5.3	0	0	0	0	0	0	0	RetM	Evely, 93, 44, no 41
413	Hel	0	4	2	5	1	0	3	1	0	0	0	1	0	0	0	0	0	0	0	0	0	0	Evely, 93, 49, no 166b
412	Hel	0	4	2	5	1	0	3	1	0	0	0	1	0	0	0	0	0	0	0	0	0	0	Evely, 93, 49, no 166a
415	KaC	0	1	2	5	1	0	3	1	0	0	0	1	17	4.9	0	0	0	0	0	0	0	2716	Evely, 93, 42, no 3a
444	Kal	0	3	2	6	1	0	3	1	0	0	0	1	16	4.2	0	0	0	0	0	0	0	1457	Evely, 93, 49, no 167
416	Kar	0	0	0	0	1	0	3	1	0	0	0	1	20	5.4	0	0	0	95	4.6	0	0.2	BriM	Evely, 93, 42, no 14
331	Kno	S House	2	2	2	1	0	3	1	0	0	0	1	16	4	0	0	0	0	0	0	0	0	Evely, 93, 42, no 7
333	Kno	UM L	2	2	2	1	3	3	1	0	0	0	1	20	5.5	0	0	1200	95	5	0	0	0	Evely, 93, 42, no 26
330	Kno	S House	2	2	2	1	0	3	1	0	0	0	1	16	3.8	0	0	0	0	0	0	0	0	Evely, 93, 42, no 6
334	Kno	NW House	2	2	2	1	0	3	1	0	0	0	1	18	4.9	0	0	1100	97	1.8	1.4	0.6	AshM	Evely, 93, 42, no 42
332	Kno	Cel SP	2	2	1	1	0	3	1	0	0	0	1	18	0	0	0	0	0	0	0	0	2268	Evely, 93, 42, no 15

Appendix V.2 Bronze objects. Sorted by object type / period / site. Total 1857

No	Site	Co	G	L	S	Pe	Da	Ot	C	TM	TK	TP	Cn	Le	Wi	He	Rd	We	Cu	Sn	As	Pb	Mu	Ref
328	Kno	NW House	2	2	2	1	0	3	1	0	0	0	1	20	5.2	0	0	0	84.8	9.6	0.3	0.2	2076	Evely, 93, 42, no 4; Mangou / Ioannou, 98, 94
329	Kno	NW House	2	2	2	1	0	3	1	0	0	0	1	18	5.5	0	0	0	72.7	6.4	0.1	0	2077	Evely, 93, 42, no 5; Mangou / Ioannou, 98, 94
338	Kno	NW House	2	2	2	1	0	3	1	0	0	0	1	12	0	0	0	0	0	0	0	0	2078	Evely, 93, 47, no 119
351	Mal	Quart. D	2	1	2	1	0	3	1	0	0	0	1	20	0	0	0	0	0	0	0	0	2232	Evely, 93, 44, no 43
341	Mal	Quart. E	2	1	2	1	0	3	1	0	0	0	1	19	5.5	0	0	0	0	0	0	0	2268	Evely, 93, 42, no 16
354	Mal	Quart. E	2	1	2	1	0	3	1	0	0	0	1	19	4.2	0	0	0	0	0	0	0	2380	Evely, 93, 47, no 132
352	Mal	Quart. Z	2	1	2	1	0	3	1	0	0	0	1	18	5.2	0	0	0	0	0	0	0	0	Evely, 93, 44, no 44
355	Mal	Quart. Z	2	1	2	1	0	3	1	0	0	0	1	0	0	0	0	0	0	0	0	0	0	Evely, 93, 49, no 148
353	Mal	Quart. Z	2	1	2	1	0	3	1	0	0	0	1	15	4	0	0	0	0	0	0	0	0	Evely, 93, 44, no 45
342	Mal	Quart. E	2	1	2	1	0	3	1	0	0	0	1	18	5.1	0	0	0	0	0	0	0	2379	Evely, 93, 42, no 17
344	Mal	Quart. Z	2	1	2	1	3	3	1	0	0	0	1	16	5.7	0	0	0	0	0	0	0	2328	Evely, 93, 42, no 28
345	Mal	Quart. Z	2	1	2	1	3	3	1	0	0	0	1	18	6.3	0	0	0	0	0	0	0	2329	Evely, 93, 42, no 29
356	Mal	0	2	1	2	1	0	3	1	0	0	0	1	0	0	0	0	0	0	0	0	0	2285a	Evely, 93, 49, no 149
343	Mal	Quart. Z	2	1	2	1	3	3	1	0	0	0	1	17	6.5	0	0	0	0	0	0	0	2327	Evely, 93, 42, no 27
346	Mal	Quart. Z	2	1	2	1	3	3	1	0	0	0	1	17	6.1	0	0	0	0	0	0	0	2330	Evely, 93, 42, no 30
349	Mal	Quart. Z	2	1	2	1	3	3	1	0	0	0	1	19	7	0	0	0	0	0	0	0	2333	Evely, 93, 44, no 33
350	Mal	Quart. Z	2	1	2	1	3	3	1	0	0	0	1	17	6.3	0	0	0	0	0	0	0	2334	Evely, 93, 44, no 34
347	Mal	Quart. Z	2	1	2	1	3	3	1	0	0	0	1	16	6.6	0	0	0	0	0	0	0	2331	Evely, 93, 42, no 31
348	Mal	Quart. Z	2	1	2	1	3	3	1	0	0	0	1	17	5.8	0	0	0	0	0	0	0	2332	Evely, 93, 44, no 32
414	Mel	0	2	2	7	1	0	3	1	0	0	0	1	0	0	0	0	0	0	0	0	0	0	Evely, 93, 44, no 45a
1745	Moc	House C3	1	1	3	1	4	3	1	0	0	0	1	0	0	0	0	0	0	0	0	0	0	Soles / Davaras, 96, 194, no CA 46
1753	Moc	House C7	1	1	3	1	4	3	1	0	0	3	0	0	0	0	0	0	0	0	0	0	0	Soles / Davaras, 96, 201, no CA 80
1752	Moc	House C7	1	1	3	1	4	3	1	0	0	3	0	0	0	0	0	0	0	0	0	0	0	Soles / Davaras, 96, 201, no CA 72
1746	Moc	House C3	1	1	3	1	4	3	1	0	0	0	1	0	0	0	0	0	0	0	0	0	0	Soles / Davaras, 96, 194, no CA 47
408	Nek	0	4	4	4	1	0	3	1	0	0	0	1	17	5.5	0	0	0	85	15	0.3	0	ChaM	Evely, 93, 42, no 18
409	Nek	Sanct.	4	2	8	1	0	3	1	0	0	0	1	20	5.4	0	0	0	85	14	0.2	0	0	Evely, 93, 42, no 19
410	NiC	0	2	1	4	1	4	3	1	0	0	0	1	14	5.1	0	0	0	0	0	0	0	2266	Evely, 93, 44, no 35
361	Pak	0	1	1	3	1	4	3	1	0	0	0	1	0	0	0	0	0	89	0.2	0.7	0.1	1372	Evely, 93, 49, no 144; Mangou / I0annou, 98, 95
362	Pak	0	1	1	3	1	4	3	1	0	0	0	1	0	0	0	0	0	87.9	0.2	1.1	0.1	1376a	Evely, 93, 49, no 145a; Mangou / Ioannou, 98, 95
1365	Pak	Block Chi	1	1	3	1	4	3	1	0	0	0	1	0	0	0	0	0	0	0	0	0	0	Dawkins, 04-05, 284
360	Pak	0	1	1	3	1	4	3	1	0	0	0	1	0	0	0	0	0	91.9	4.7	0.4	0.1	852	Evely, 93, 49, no 143; Mangou / Ioannou, 98, 94
358	Pak	0	1	1	3	1	0	3	1	0	0	0	1	16	6.2	0	0	0	85.7	6.3	0.5	0	1379	Evely, 93, 47, no 136; Mangou / Ioannou, 98, 95
366	Pak	0	1	1	3	1	4	3	1	0	0	0	1	0	0	0	0	0	0	0	0	0	1376b	Evely, 93, 49, no 145b
359	Pak	0	1	1	3	1	4	3	1	0	0	0	1	0	0	0	0	0	86.4	5.5	0.3	0.2	1378	Evely, 93, 49, no 142; Mangou / Ioannou, 98, 95
1362	Pak	Block Chi	1	1	3	1	4	3	1	0	0	0	1	0	0	0	0	0	0	0	0	0	0	Dawkins, 04-05, 282
1355	Pak	Block Chi	1	1	3	1	4	3	1	0	0	0	1	0	0	0	0	0	0	0	0	0	0	Dawkins, 04-05, 282
443	Pet	0	1	2	6	1	0	3	1	0	0	0	1	0	0	0	0	0	0	0	0	0	0	Evely, 93, 49, no 156
322	Pha	63d	3	2	1	1	4	3	1	0	0	0	1	16	4.5	0	0	0	95	2.6	0.4	0.2	349	Evely, 93, 44, no 54; Mangou / I0annou, 98, 94
321	Pha	63d	3	2	1	1	4	3	1	0	0	0	1	15	5	0	0	0	97	0.2	0.6	0.1	348	Evely, 93, 44, no 53; Mangou / Ioannou, 98, 94
314	Pha	63d	3	2	1	1	4	3	1	0	0	0	1	24	6	0	0	0	0	0	0	0	341	Evely, 93, 44, no 46

Appendix V.2 Bronze objects. Sorted by object type / period / site. Total 1857

No	Site	Co	G	L	S	Pe	Da	Ot	C	TM	TK	TP	Cn	Le	Wi	He	Rd	We	Cu	Sn	As	Pb	Mu	Ref
315	Pha	63d	3	2	1	1	4	3	1	0	0	0	2	20	6	0	0	0	89	5.2	0.2	0.1	342	Evely, 93, 44, no 47; Mangou / Ioannou, 98, 94
313	Pha	XLII	3	2	1	1	1	3	1	0	0	0	1	10	3.5	0	0	0	0	0	0	0	1771	Evely, 93, 41, no 3
316	Pha	63d	3	2	1	1	4	3	1	0	0	0	1	20	5	0	0	0	91.4	3.6	0.5	0.2	343	Evely, 93, 44, no 48; Mangou / Ioannou, 98, 94
320	Pha	63d	3	2	1	1	4	3	1	0	0	0	1	18	5	0	0	0	89.6	3.7	0.2	0.1	347	Evely, 93, 44, no 52; Mangou / Ioannou, 98, 94
318	Pha	63d	3	2	1	1	4	3	1	0	0	0	1	18	6	0	0	0	88.2	6.7	0.1	0.1	345	Evely, 93, 44, no 50; Mangou / Ioannou, 98, 94
317	Pha	63d	3	2	1	1	4	3	1	0	0	0	1	22	6	0	0	0	88.2	6	0.2	0.2	344	Evely, 93, 44, no 49; Mangou / Ioannou,98, 94
319	Pha	63d	3	2	1	1	4	3	1	0	0	0	1	19	5.4	0	0	0	96	1.6	0.5	0.3	346	Evely, 93, 44, no 51; Mangou / Ioannou, 98, 94
393	Psc	0	1	1	3	1	0	3	1	0	0	0	1	17	4.7	0	0	0	0	0	0	0	4212	Evely, 93, 44, no 55
1573	Sel	0	1	2	0	1	0	3	1	0	0	0	1	20	5.3	0	0	1200	919	6.8	0.5	0.1	AshM	Evely, 93, 46, no 96; Northover / Evely,,95, 87, no 4, 95
1823	Tyl	House A	4	2	4	1	4	3	1	0	0	0	1	0	0	0	0	0	0	0	0	0	0	Driessen / Macdonald, 97, 129
1822	Tyl	House A	4	2	4	1	4	3	1	0	0	0	1	0	0	0	0	0	0	0	0	0	0	Driessen / Macdonald, 97, 129
1818	Tyl	House A	4	2	4	1	4	3	1	0	0	0	1	0	0	0	0	0	0	0	0	0	0	Driessen / Macdonald, 97, 129
417	Vor	0	1	0	0	1	0	3	1	0	0	0	1	20	5.5	0	0	0	0	0	0	0	2504	Evely, 93, 42, no 20
35	Zak	XXIV	1	1	1	1	4	3	1	0	0	0	1	18	4.5	0	0	1000	0	0	0	0	2614	Platon, 88, II, 200, no 2
39	Zak	XV	1	1	1	1	4	3	1	0	0	0	1	14	3.7	0	0	490	0	0	0	0	2596	Platon, 88, II, 201, no 6
34	Zak	House D	1	1	2	0	0	3	1	0	0	0	3	8	5.4	0	0	360	0	0	0	0	0	Platon, 88, II, 200, no 1
38	Zak	XXV	1	1	1	1	4	3	1	0	0	0	1	13	5.5	0	0	400	0	0	0	0	2615	Platon, 88, II, 201, no 5
40	Zak	IV	1	1	1	1	4	3	1	0	0	0	1	16	5.1	0	0	1000	0	0	0	0	2595	Platon, 88, II, 202, no 7
36	Zak	XXIV	1	1	1	1	4	3	1	0	0	0	2	17	4.3	0	0	900	0	0	0	0	2619	Platon, 88, II, 200, no 3
42	Zak	House C	1	1	2	1	0	3	1	0	0	0	1	22	5.8	0	0	1325	0	0	0	0	652	Platon, 88, II, 202, no 9
41	Zak	IV	1	1	1	1	4	3	1	0	0	0	1	16	4.4	0	0	900	0	0	0	0	2594	Platon, 88, II, 202, no 8
43	Zak	House C	1	1	2	1	0	3	1	0	0	0	1	20	5.6	0	0	1380	0	0	0	0	653	Platon, 88, II, 203, no 10
340	Iso	Tomb 2	2	2	6	2	5	3	1	0	0	0	1	20	4.2	0	0	0	0	0	0	0	1751	Evely, 93, 47, no 135
1171	Iso	Tomb 2	2	2	6	2	5	3	1	0	0	0	1	20	0	0	0	0	0	0	0	0	0	Evans, 14, 58, no 2a
339	Kno	UM H	2	2	2	2	5	3	1	0	0	0	3	9	3.4	0	0	340	99	1	0	0	0	Evely, 93, 47, no 134
446	Arm	Tomb 19	4	2	6	3	8	3	1	0	0	0	1	0	0	0	0	0	0	0	0	0	0	Evely, 93, 49, no 170
438	Sam	0	4	2	5	3	8	3	1	0	0	0	1	16	4.1	0	0	0	0	0	0	0	ChaM	Evely, 93, 49, no 140,;Kanta, 80, 236
437	Sam	0	4	2	5	3	8	3	1	0	0	0	1	20	4.2	0	0	0	0	0	0	0	ChaM	Evely, 93, 49, no 139; Kanta, 80, 236
439	Ste	0	1	2	4	3	8	3	1	0	0	0	1	0	0	0	0	0	0	0	0	0	0	Evely, 93, 49, no 146
478	0	0	0	0	0	0	0	5	1	0	0	0	1	33	7	0	0	0	0	0	0	0	1712	Evely, 93, 63, no 15
477	0	0	0	0	0	0	0	5	1	0	0	0	1	33	5.8	0	0	0	0	0	0	0	1712b	Evely, 93, 63, no 14
468	HaT	Room 16	3	2	4	0	0	5	1	0	0	0	1	13	4.8	0	0	0	90.7	0.2	0.3	0	1223	Evely, 93, 63, no 5; Mangou / Ioannou, 98, 94
474	Kno	Hou NB	2	2	2	0	0	5	1	0	0	0	1	31	5.6	0	0	1500	99	0	0.5	0	AshM	Evely, 93, 63, no12; Northover / Evely, 95, 87, no 7
475	Kno	0	2	2	2	0	0	5	1	0	0	0	1	30	5.2	0	0	1500	98	0.7	0.2	0.6	AshM	Evely, 93, 63, no 13
479	Mal	0	2	1	0	0	0	5	1	0	0	0	1	0	0	0	0	0	0	0	0	0	0	Evely, 93, 63, no 17
470	Kno	NW House	2	2	2	1	0	5	1	0	0	0	1	30	4.5	0	0	0	0	0	0	0	0	Evely, 93, 63, no 8
472	Kno	NW House	2	2	2	1	0	5	1	0	0	0	1	30	0	0	0	0	0	0	0	0	0	Evely, 93, 63, no 10
471	Kno	NW House	2	2	2	1	0	5	1	0	0	0	1	30	5.5	0	0	0	0	0	0	0	0	Evely, 93, 63, no 9
469	Kno	NW House	2	2	2	1	0	5	1	0	0	0	1	34	5.5	0	0	0	0	0	0	0	0	Evely, 93, 63, no 7
473	Kno	NW House	2	2	2	1	0	5	1	0	0	0	1	39	0	0	0	0	0	0	0	0	0	Evely, 93, 63, no 11

Appendix V.2 / 14

Appendix V.2 Bronze objects. Sorted by object type / period / site. Total 1857

No	Site	Co	G	L	S	Pe	Da	Ot	C	TM	TK	TP	Cn	Le	Wi	He	Rd	We	Cu	Sn	As	Pb	Mu	Ref
58	Zak	House G	1	1	2	1	0	5	1	0	0	0	1	25	5.4	0	0	1280	0	0	0	0	663	Platon, 88, II, 208, no 5
55	Zak	XXIV	1	1	1	1	4	5	1	0	0	0	3	13	4.9	0	0	630	0	0	0	0	2618	Platon, 88, II, 207, no 2
57	Zak	House A	1	1	2	1	0	5	1	0	0	0	1	21	5.5	0	0	1000	0	0	0	0	664	Platon, 88, II, 208, no 4
476	Lou	0	4	2	4	3	0	5	1	0	0	0	1	13	4	0	0	0	0	0	0	0	3865	Evely, 93, 63, no 6
489	0	0	0	0	0	0	0	6	1	0	0	0	1	15	5.3	0	0	0	0	0	0	0	400	Evely, 93, 68, no 11
486	Mil	0	1	1	0	0	0	6	1	0	0	0	1	13	3.7	0	0	300	91	7.9	0.8	0.4	AshM	Evely, 93, 68, no 8; Northover / Evely, 95, 87, no 9
487	RoE	0	1	2	0	0	0	6	1	0	0	0	1	16	4.2	0	0	0	0	0	0	0	1023	Evely, 93, 68, no 9
488	Ark	0	2	2	0	1	0	6	1	0	0	0	1	0	0	0	0	0	0	0	0	0	0	Evely, 93, 68, no 10
480	Moc	0	1	1	3	1	0	6	1	0	0	0	1	15	3.7	0	0	300	87	6	0	6.9	AshM	Evely, 93, 67, no 2
481	Moc	0	1	1	3	1	0	6	1	0	0	0	1	16	3.6	0	0	325	89	7.5	0	3.7	AshM	Evely, 93, 68, no 6; Northover / Evely, 95, 87, no 6
483	Pak	0	1	1	0	1	0	6	1	0	0	0	1	14	4	0	0	0	0	0	0	0	0	Evely, 93, 68, no 4
484	Pha	0	1	2	1	0	0	6	1	0	0	0	1	18	3.9	0	0	0	0	0	0	0	1308	Evely, 93, 68, no 5
482	Vai	0	1	1	4	0	3	6	1	0	0	0	1	19	4.3	0	0	0	0	0	0	0	0	Evely, 93, 68, no 3
485	Kav	0	1	2	0	3	0	6	1	0	0	0	1	14	3.8	0	0	0	0	0	0	0	2345	Evely, 93, 68, no 7
494	Pak	0	1	1	3	1	0	7	1	0	0	0	1	15	5.4	0	0	0	0	0	0	0	1130	Evely, 93, 72, no 1
497	Vai	0	1	1	4	1	3	7	1	0	0	0	1	0	0	0	0	0	0	0	0	0	0	Evely, 93, 72, no 13
37	Zak	XXIV	1	1	1	1	4	7	1	0	0	0	1	16	5.4	0	0	500	0	0	0	0	2620	Platon, 88, II, 201, no 4
496	EkP	Hag. Apos	2	2	6	2	6	7	1	0	0	0	1	0	0	0	0	0	0	0	0	0	0	Evely, 93, 72, no 12
495	Olo	Tomb 36	1	1	6	3	8	7	1	0	0	0	1	5	1.1	0	0	0	0	0	0	0	0	Evely, 93, 72, no 6
490	Rog	0	4	2	0	0	0	8	1	0	0	0	1	15	3.5	0	0	0	0	0	0	0	538	Evely, 93, 71, no 3
491	HaT	0	3	2	4	1	0	8	1	0	0	0	1	38	5.4	0	0	0	0	0	0	0	1221	Evely, 93, 71, no 4
492	HaT	0	3	2	4	1	0	8	1	0	0	0	1	25	5.6	0	0	0	0	0	0	0	1222	Evely, 93, 71, no 5
493	Ita	0	1	1	4	1	3	8	1	0	0	0	1	0	0	0	0	0	0	0	0	0	0	Evely, 93, 71, no 6
56	Zak	XXIV	1	1	1	1	4	8	1	0	0	0	1	13	4.5	0	0	700	0	0	0	0	2616	Platon, 88, II, 208, no 3
54	Zak	XXIV	1	1	1	1	4	8	1	0	0	0	1	17	4.6	0	0	1170	0	0	0	0	2617	Platon, 88, II, 207, no 1
1444	Kma	0	2	2	5	3	0	8	1	0	0	0	2	0	0	0	0	0	0	0	0	0	0	Kanta, 80, 164, 183
309	0	0	4	0	0	0	0	9	1	0	0	0	2	25	5	0	0	0	0	0	0	0	ChaM	Evely, 93, 31, no 33
312	0	0	0	0	0	0	0	9	1	0	0	0	1	0	0	0	0	0	0	0	0	0	GiaC	Evely, 93, 33, no 76
306	Apd	0	4	2	5	0	0	9	1	0	0	0	1	2.5	2	0	0	0	0	0	0	0	0	Evely, 93, 33, no 60
305	Apd	0	4	2	5	0	0	9	1	0	0	0	1	2.5	2	0	0	0	0	0	0	0	0	Evely, 93, 33, no 59
308	Apd	0	4	2	5	0	0	9	1	0	0	0	1	2.5	2	0	0	0	0	0	0	0	0	Evely, 93, 33, no 62
307	Apd	0	4	2	5	0	0	9	1	0	0	0	1	2.5	2	0	0	0	0	0	0	0	0	Evely, 93, 33, no 61
304	Apd	0	4	2	5	0	0	9	1	0	0	0	1	2.5	2	0	0	0	0	0	0	0	0	Evely, 93, 33, no 58
299	HaT	0	3	2	4	0	0	9	1	0	0	0	1	0	0	0	0	0	0	0	0	0	1541	Evely, 93, 33, no 50
300	HaT	Palace	3	2	4	0	0	9	1	0	0	0	2	0	0	0	0	0	0	0	0	0	0	Evely, 93, 33, no 77
298	HaT	Tomb	3	2	6	0	0	9	1	0	0	0	1	52	6.5	0	0	0	0	0	0	0	1531	Evely, 93, 33, no 49
289	Kno	0	2	2	2	0	0	9	1	0	0	0	1	9.2	2.8	0	0	0	0	0	0	0	StraM	Evely, 93, 33, no 57
282	Kno	0	2	2	2	0	0	9	0	0	0	0	2	60	7.5	0	0	0	0	0	0	0	StraM	Evely, 93, 31, no 22
280	Pak	Town	1	1	3	0	0	9	1	0	0	0	2	5	0	0	0	0	0	0	0	0	0	Evely, 93, 26, no 5
281	Pak	Town	1	1	3	0	0	9	1	0	0	0	2	9	0	0	0	0	0	0	0	0	0	Evely, 93, 26, no 6

Appendix V.2 Bronze objects. Sorted by object type / period / site. Total 1857

No	Site	Co	G	L	S	Pe	Da	Ot	C	TM	TK	TP	Cn	Le	Wi	He	Rd	We	Cu	Sn	As	Pb	Mu	Ref
310	Pse	0	1	1	3	0	0	9	1	0	0	0	1	0	0	0	0	0	0	0	0	0	1480	Evely, 93, 33, no 74
311	Rou	0	3	2	9	0	0	9	1	0	0	0	1	0	0	0	0	0	0	0	0	0	2484	Evely, 93, 33, no 75
272	Gou	E 16	1	1	3	1	0	9	1	0	0	0	2	5.7	1.3	0	0	0	0	0	0	0	602	Evely, 93, 31, no 17
278	Gou	0	1	1	3	1	0	9	1	0	0	0	1	36	5	0	0	0	0	0	0	0	0	Evely, 93, 31, no 37
274	Gou	0	1	1	3	1	0	9	1	0	0	0	1	40	6	0	0	0	0	0	0	0	0	Evely, 93, 31, no 21
275	Gou	F 18	1	1	3	1	0	9	1	0	0	0	1	45	6	0	0	0	0	0	0	0	570	Evely, 93, 31, no 34
270	Gou	E 16	1	1	3	1	0	9	1	0	0	0	2	5.4	1	0	0	0	0	0	0	0	0	Evely, 93, 26, no 7
277	Gou	0	1	1	3	1	0	9	1	0	0	0	2	28	5.5	0	0	0	0	0	0	0	0	Evely, 93, 31, no 36
271	Gou	E 16	1	1	3	1	0	9	1	0	0	0	2	5.8	1	0	0	0	0	0	0	0	578	Evely, 93, 26, no 8
276	Gou	Gi	1	1	3	1	0	9	1	0	0	0	2	56	8.3	0	0	0	0	0	0	0	572	Evely, 93, 31, no 35
273	Gou	Hill House	1	1	3	1	0	9	1	0	0	0	1	29	8	0	0	0	0	0	0	0	571	Evely, 93, 31, no 20
297	HaT	Palace	3	1	4	1	0	9	1	0	0	0	2	51	13	0	0	700	0	0	0	0	702	Evely, 93, 33, no 46
296	HaT	Palace	3	2	4	1	0	9	1	0	0	0	1	144	12	0	0	0	92	0.8	0.6	0	701	Evely, 93, 31, no 23; Mangou / Ioannou, 98, 94
286	Kno	S House	2	2	2	1	0	9	1	0	0	0	2	53	9	0	0	0	0	0	0	0	0	Evely, 93, 31, no 41
283	Kno	Hou NB	2	2	2	1	0	9	1	0	0	0	2	42	8.5	0	0	350	0	0	0	0.1	AshM	Evely, 93, 31, no 38
288	Kno	S House	2	2	2	1	0	9	1	0	0	0	1	163	19	0	0	0	74.4	0.2	0.1	0	2053	Evely, 93, 33, no 45; Mangou / Ioannou, 98, 94
284	Kno	S House	2	2	2	1	0	9	1	0	0	0	1	37	8	0	0	0	0	0	0	0	0	Evely, 93, 31, no 39
285	Kno	S House	2	2	2	1	0	9	1	0	0	0	1	45	6.5	0	0	0	0	0	0	0	0	Evely, 93, 31, no 40
293	Mal	Quart. L	2	1	2	1	0	9	1	0	0	0	2	4	1.8	0	0	0	0	0	0	0	1963	Evely, 93, 31, no 16
294	Mal	House E	2	1	2	1	0	9	1	0	0	0	2	7.8	3	0	0	0	0	0	0	0	0	Evely, 93, 31, no 18
1802	Mal	Quart.D, Db	2		2	1	0	9	1	0	0	0	1	0	0	0	0	0	0	0	0	0	0	Driessen / Macdonald, 97, 186
295	Mal	Quart. Z	2		2	1	0	9	1	0	0	0	1	141	16	0	0	0	0	0	0	0	2191	Evely, 93, 33, no 44
279	Pak	Town	1	3	3	4	0	9	1	0	0	0	2	5.7	1.5	0	0	0	0	0	0	0	0	Evely, 93, 26, no 4
303	Tyl	0	4	2	4	0	0	9	1	0	0	0	1	9.5	3	0	0	0	0	0	0	0	0	Evely, 93, 33, no 51
3	Zak	XXVIII	1	1	1	4	0	9	1	0	0	0	1	145	17.7	0	0	4919	0	0	0	0	2673	Platon, 88, II, 190, no 3
92	Zak	House I	1	1	2	0	0	9	1	0	0	0	1	6.7	4	0	0	20	0	0	0	0	657	Platon, 88, II, 217, no 4/2
95	Zak	House I	1	1	2	1	0	9	1	0	0	0	1	6.8	4.2	0	0	20	0	0	0	0	657	Platon, 88, II, 217, no 4/5
93	Zak	House I	1	1	2	0	0	9	1	0	0	0	1	6.7	4	0	0	20	0	0	0	0	657	Platon, 88, II, 217, no 4/3
91	Zak	House I	1	1	2	0	0	9	1	0	0	0	1	6.9	3.9	0	0	20	0	0	0	0	657	Platon, 88, II, 217, no 4/1
9	Zak	XXIV	1	1	1	4	0	9	1	0	0	0	3	52	9.3	0	0	1540	0	0	0	0	6199	Platon, 88, II, 193, no 9
88	Zak	House B	1	1	2	3	0	9	1	0	0	0	2	4.9	1.9	0	0	0	0	0	0	0	0	Platon, 88, II, 217, no 1
5	Zak	XXVIII	1	1	1	4	0	9	1	0	0	0	1	168	21	0	0	7328	0	0	0	0	2675	Platon, 88, II, 191, no 5
94	Zak	House I	1	1	2	0	0	9	1	0	0	0	1	6.5	4.1	0	0	20	0	0	0	0	657	Platon, 88, II, 217, no 4/4
1	Zak	XXIV	1	1	1	4	0	9	1	0	0	0	2	146	12.5	0	0	0	0	0	0	0	2613	Platon, 88, II, 190, no 1
6	Zak	XXVIII	1	1	1	4	0	9	1	0	0	0	1	154	21.5	0	0	0	0	0	0	0	2613	Platon, 88, II, 192, no 6
89	Zak	Buil. Ni	1	1	2	1	0	9	1	0	0	0	3	5.9	1.7	0	0	0	0	0	0	0	0	Platon, 88, II, 217, no 2
4	Zak	XXVIII	1	1	1	4	0	9	1	0	0	0	1	142	12.7	0	0	0	0	0	0	0	2674	Platon, 88, II, 191, no 4
2	Zak	X	1	1	1	4	0	9	1	0	0	0	2	105	12.8	0	0	2460	0	0	0	0	2598	Platon, 88, II, 190, no 2
7	Zak	X	1	1	1	4	0	9	1	0	0	0	2	142	10.6	0	0	0	0	0	0	0	2599	Platon, 88, II, 192, no 7
8	Zak	X	1	1	1	1	0	9	1	0	0	0	1	120	9.3	0	0	0	0	0	0	0	2600	Platon, 88, II, 192, no 8

Appendix V.2 Bronze objects. Sorted by object type / period / site. Total 1857

No	Site	Co	G	L	S	Pe	Da	Ot	C	TM	TK	TP	Cn	Le	Wi	He	Rd	We	Cu	Sn	As	Pb	Mu	Ref
90	Zak	House Ra	1	1	2	1	3	9	0	0	0	0	3	5	1.2	0	0	0	0	0	0	0	0	Platon, 88, II, 217, no 3
292	Kno	UM M	2	2	2	2	5	9	0	0	0	0	2	5	6	0	0	0	0	20	0	0	0	Evely, 93, 33, no 72
290	Kno	UM M	2	2	2	2	5	9	0	0	0	0	2	7.5	8.8	0	0	0	0	0	0	0	0	Evely, 93, 33, no 70
291	Kno	UM M	2	2	2	2	5	9	0	0	0	0	2	12	7.5	0	0	0	0	7	0	0	0	Evely, 93, 33, no 71
287	Kno	UM M	2	2	2	2	5	9	0	0	0	0	2	17	8.5	0	0	0	0	0	0	0	0	Evely, 93, 31, no 42
302	Psy	0	1	2	7	3	0	9	0	0	0	0	2	3.9	0.7	0	0	0	0	0	0	0	AshM	Evely, 93, 31, no 9
301	ZaP	Tomb 33	2	2	6	3	0	9	0	0	0	0	1	48	5.3	0	0	0	0	0	0	0	0	Evely, 93, 31, no 43
68	Zak	XXIV	1	1	1	1	4	10	0	0	0	0	1	23	0	0	0	0	0	0	0	0	2627	Platon, 88, II, 211, no 5
62	Zak	XXIV	1	1	1	1	4	10	0	0	0	0	1	9	0.8	0	0	20	0	0	0	0	0	Platon, 88, II, 210, no 1
65	Zak	IV	1	1	1	1	4	10	0	0	0	0	1	11	0	0	0	0	0	0	0	0	0	Platon, 88, II, 211, no 2
69	Zak	House I	1	1	2	1	0	10	0	0	0	0	1	3.5	0	0	0	65	0	0	0	0	658a	Platon, 88, II, 212, no 6
64	Zak	IV	1	1	1	1	4	10	0	0	0	0	1	13	0	0	0	0	0	0	0	0	0	Platon, 88, II, 211, no 1
66	Zak	XXVIII	1	1	1	1	4	10	0	0	0	0	3	11	0	0	0	0	0	0	0	0	0	Platon, 88, II, 211, no 3
63	Zak	X	1	1	1	1	4	10	0	0	0	0	1	12	1	0	0	0	0	0	0	0	3134	Platon, 88, II, 210, no 2
67	Zak	XXVIII	1	1	1	1	4	10	0	0	0	0	1	11	0	0	0	0	0	0	0	0	2626	Platon, 88, II, 211, no 4
577	Arh	0	2	2	0	0	0	11	0	0	0	0	1	0	0	0	0	0	0	0	0	0	0	Evely, 93, 101, no 7
574	Rou	0	3	2	5	0	0	11	0	0	0	0	1	0	0	0	0	0	0	0	0	0	0	Evely, 93, 101, no 4
578	Vav	0	0	0	0	0	0	11	0	0	0	0	1	0	0	0	0	0	0	0	0	0	2455	Evely, 93, 101, no 8
575	Zir	0	1	2	5	0	0	11	0	0	0	0	1	0	0	0	0	0	0	0	0	0	0	Evely, 93, 101, no 5
576	Zir	0	1	2	5	0	0	11	0	0	0	0	1	0	0	0	0	0	0	0	0	0	0	Evely, 93, 101, no 6
572	Gou	0	1	1	3	1	0	11	0	0	0	0	1	12	5	0	0	0	0	0	0	0	965	Evely, 93, 101, no 18
567	HaT	Room 55	3	2	4	1	0	11	0	0	0	0	1	0	0	0	0	0	0	0	0	0	0	Evely, 93, 101, no 10a
566	HaT	Room 55	3	2	4	1	0	11	0	0	0	0	1	0	0	0	0	0	0	0	0	0	0	Evely, 93, 101, no 10
568	HaT	Room 37	3	2	4	1	0	11	0	0	0	0	1	23	8	0	0	7240	0	0	0	0	831	Evely, 93, 101, no 13
569	HaT	Room 37	3	2	4	1	0	11	0	0	0	0	1	14	7.5	0	0	4160	92.5	1.2	0.6	0	1253	Evely, 93, 101, no 14; Mangou / Ioannou, 98, 94
570	Kno	S House	2	2	2	1	0	11	0	0	0	0	1	11	2.5	0	0	0	0	0	0	0	0	Evely, 93, 101, no 11
565	Mal	House Zb	2	1	2	1	0	11	0	0	0	0	1	15	2	0	0	0	0	0	0	0	0	Evely, 93, 97, no 1
573	Psy	0	1	2	7	1	0	11	0	0	0	0	1	10	4.4	0	0	0	0	0	0	0	1379	Evely, 93, 101, no 19
571	Psy	0	1	2	7	1	0	11	0	0	0	0	2	6.6	3.8	0	0	600	77	14	1.2	6.9	AshM	Evely, 93, 101, no 2; Northover / Evely, 95, 87, no 10
45	Zak	IV	1	1	1	1	4	11	0	0	0	0	1	9.2	5.4	0	0	850	0	0	0	0	2593	Platon, 88, II, 203
1395	Lou	0	4	2	4	3	0	11	0	0	0	0	1	0	0	0	0	0	0	0	0	0	0	Kanta, 80, 23
534	Kno	Roy. Road	2	2	2	0	0	12	0	0	0	0	2	5.6	0.5	0	0	0	0	0	0	0	0	Evely, 93, 90, no 79
524	Kno	0	2	2	0	0	0	12	0	0	0	0	1	4.7	0.4	0	0	0	0	0	0	0	StraM	Evely, 93, 90, no 63
526	Kno	Hog Hou	2	2	2	0	0	12	0	0	0	0	1	0	0	0	0	0	0	0	0	0	0	Evely, 93, 90, no 65+
546	Kno	Roy. Road	2	2	2	0	0	12	0	0	0	0	1	0	0	0	0	0	0	0	0	0	0	Evely, 93, 92, no 131
525	Kno	0	2	2	0	0	0	12	0	0	0	0	1	0	0	0	0	0	0	0	0	0	StraM	Evely, 93, 90, no 64+
540	Kno	0	2	2	0	0	0	12	0	0	0	0	1	11	0.5	0	0	0	0	0	0	0	StraM	Evely, 93, 92, no 103
518	Kno	0	2	2	1	0	0	12	0	0	0	0	1	13	0.7	0	0	0	0	0	0	0	0	Evely, 93, 88, no 26
547	Kno	Roy. Road	2	2	2	0	0	12	0	0	0	0	1	0	0	0	0	0	0	0	0	0	0	Evely, 93, 92, no 132
504	Mal	House E	2	1	2	0	0	12	0	0	0	0	1	0	0	0	0	0	0	0	0	0	2441	Evely, 93, 92, no 110

Appendix V.2 Bronze objects. Sorted by object type / period / site. Total 1857

No	Site	Co	G	L	S	Pe	Da	Ot	C	TM	TK	TP	Cn	Le	Wi	He	Rd	We	Cu	Sn	As	Pb	Mu	Ref
503	Mal	House E	2	1	2	0	0	12	1	0	0	0	1	22	1.5	0	0	0	0	0	0	0	0	Evely, 93, 90, no 85
512	Pak	LM dep	1	1	3	0	0	12	1	0	0	0	1	0	0	0	0	0	0	0	0	0	0	Evely, 93, 88, no 16
510	Pak	LM dep	1	1	3	0	0	12	1	0	0	0	1	0	0	0	0	0	0	0	0	0	0	Evely, 93, 88, no 14
513	Pak	LM dep	1	1	3	0	0	12	1	0	0	0	1	0	0	0	0	0	0	0	0	0	0	Evely, 93, 88, no 17
511	Pak	LM dep	1	1	3	0	0	12	1	0	0	0	1	0	0	0	0	0	0	0	0	0	0	Evely, 93, 88, no 15
515	Pak	Block N	1	1	3	0	0	12	1	0	0	0	1	3.5	0.2	0	0	0	0	0	0	0	0	Evely, 93, 88, no 19
514	Pak	LM dep	1	1	3	0	0	12	1	0	0	0	1	0	0	0	0	0	0	0	0	0	0	Evely, 93, 88, no 18
556	Pha	0	3	2	1	0	0	12	1	0	0	0	1	12	0.9	0	0	0	0	0	0	0	0	Evely, 93, 90, no 81
563	Psc	0	1	1	3	0	0	12	1	0	0	0	1	0	0	0	0	0	0	0	0	0	SciC	Evely, 93, 90, no 104
564	Psc	0	1	1	3	0	0	12	1	0	0	0	1	0	0	0	0	0	0	0	0	0	SciC	Evely, 93, 90, no 105
553	Gou	0	1	1	3	1	0	12	1	0	0	0	1	13	0.6	0	0	0	0	0	0	0	0	Evely, 93, 90, no 80
552	Gou	B 9	1	1	3	1	0	12	1	0	0	0	1	11	0.4	0	0	0	0	0	0	0	569	Evely, 93, 90, no 40
551	Gou	B 9	1	1	3	1	0	12	1	0	0	0	1	17	0.8	0	0	0	0	0	0	0	957	Evely, 93, 88, no 22
1666	Gou	B 9	1	1	3	1	4	12	1	0	0	0	1	17	0	0	0	0	0	0	0	0	957	Boyd Hawes, 08, 34, no 17
535	Kno	Roy. Road	2	2	2	1	4	12	1	0	0	0	1	3.9	0.4	0	0	0	0	0	0	0	0	Evely, 93, 92, no 95
538	Kno	Roy. Road	2	2	2	1	4	12	1	0	0	0	1	3.8	0.3	0	0	0	0	0	0	0	0	Evely, 93, 92, no 101
541	Kno	Roy. Road	2	2	2	1	1	12	1	0	0	0	0	0	0	0	0	0	0	0	0	0	0	Evely, 93, 92, no 119
520	Kno	Roy. Road	2	2	2	1	2	12	1	0	0	0	1	10	0.4	0	0	0	0	0	0	0	0	Evely, 93, 90, no 39
522	Kno	RR IvoD	2	2	2	1	4	12	1	0	0	0	3	3.2	0.7	0	0	0	0	0	0	0	0	Evely, 93, 90, no 60
545	Kno	Roy. Road	2	2	2	1	4	12	1	0	0	0	1	0	0	0	0	0	0	0	0	0	0	Evely, 93, 92, no 129
539	Kno	Roy. Road	2	2	2	1	4	12	1	0	0	0	3	2.4	0.3	0	0	0	0	0	0	0	0	Evely, 93, 92, no 102
543	Kno	RR IvoD	2	2	2	1	4	12	1	0	0	0	1	0	0	0	0	0	0	0	0	0	0	Evely, 93, 92, no 127
542	Kno	RR IvoD	2	2	2	1	4	12	1	0	0	0	1	0	0	0	0	0	0	0	0	0	0	Evely, 93, 92, no 126
544	Kno	RR IvoD	2	2	2	1	4	12	1	0	0	0	1	0	0	0	0	0	0	0	0	0	0	Evely, 93, 92, no 128
523	Kno	RR IvoD	2	2	2	1	4	12	1	0	0	0	1	2.2	0.4	0	0	0	0	0	0	0	0	Evely, 93, 90, no 61
521	Kno	RR IvoD	2	2	2	1	4	12	1	0	0	0	3	1.5	0.2	0	0	0	0	0	0	0	0	Evely, 93, 90, no 59
505	Mal	Quart. L	2	2	2	0	0	12	1	0	0	0	1	0	0	0	0	0	0	0	0	0	0	Evely, 93, 92, no 120
502	Mal	XXV 2	2	2	1	1	0	12	1	0	0	0	1	27	1	0	0	125	0	0	0	0	2103	Evely, 93, 90, no 84
508	Mal	Quart. L	2	2	2	1	0	12	1	0	0	0	1	0	0	0	0	0	0	0	0	0	0	Evely, 93, 92, no 123
506	Mal	Quatr. L	2	2	2	1	0	12	1	0	0	0	1	1.4	0	0	0	0	0	0	0	0	0	Evely, 93, 92, no 121
500	Mal	Quart. Z	2	2	2	1	3	12	1	0	0	0	1	12	0.4	0	0	0	0	0	0	0	0	Evely, 93, 88, no 24
509	Mal	XXV 2	2	1	1	1	0	12	1	0	0	0	2	13	1	0	0	0	0	0	0	0	0	Evely, 93, 92, no 124
499	Mal	Quart. Z	2	2	2	1	3	12	1	0	0	0	2	9.6	0.4	0	0	0	0	0	0	0	0	Evely, 93, 88, no 13
498	Mal	IV	2	1	1	1	0	12	1	0	0	0	1	5.5	0.3	0	0	0	0	0	0	0	0	Evely, 93, 88, no 12
501	Mal	House E	2	2	2	1	0	12	1	0	0	0	1	11	0.8	0	0	0	0	0	0	0	0	Evely, 93, 88, no 25
507	Mal	Quart. L	2	2	2	1	0	12	1	0	0	0	1	6.5	0	0	0	0	0	0	0	0	2387	Evely, 93, 92, no 122
1732	Moc	House A2	1	1	3	1	4	12	1	0	0	0	1	0	0	0	0	0	0	0	0	0	0	Soles / Davaras, 94, 419, no CA 23
1761	Moc	Chalinom.	1	1	3	1	4	12	1	0	0	0	1	0	0	0	0	0	0	0	0	0	0	Soles / Davaras, 96, 210, no CA 84
516	Pak	Block K	1	1	3	1	0	12	1	0	0	0	1	20	0.7	0	0	0	0	0	0	0	0	Evely, 93, 88, no 23
517	Pak	House A	1	1	3	1	0	12	1	0	0	0	1	0	0	0	0	0	0	0	0	0	0	Evely, 93, 92, no 125

Appendix V.2 Bronze objects. Sorted by object type / period / site. Total 1857

No	Site	Co	G	L	S	Pe	Da	Ot	C	TM	TK	TP	Cn	Le	Wi	He	Rd	We	Cu	Sn	As	Pb	Mu	Ref
557	Pha	0	3	2	1	1	0	12	1	0	0	0	1	14	1.2	0	0	0	0	0	0	0	0	Evely, 93, 90, no 82
559	Pha	0	3	2	1	1	0	12	1	0	0	0	1	8	0.6	0	0	0	0	0	0	0	0	Evely, 93, 90, no 94
558	Pha	0	3	2	1	1	0	12	1	0	0	0	1	14	1	0	0	0	0	0	0	0	0	Evely, 93, 90, no 83
560	Pha	0	3	2	1	1	0	12	1	0	0	0	1	12	0.4	0	0	0	0	0	0	0	0	Evely, 93, 90, no 98
554	Tyl	0	4	2	4	1	0	12	1	0	0	0	1	7	0.3	0	0	0	0	0	0	0	0	Evely, 93, 90, no 35
555	Tyl	0	4	2	4	1	0	12	1	0	0	0	1	8	0.2	0	0	0	0	0	0	0	0	Evely, 93, 90, no 36
103	Zak	House D	1	1	2	1	0	12	1	0	0	0	1	7.6	0.3	0	0	0	0	0	0	0	0	Platon, 88, II, 219, no 8
102	Zak	House D	1	1	2	1	0	12	1	0	0	0	1	5.2	0.5	0	0	0	0	0	0	0	0	Platon, 88, II, 219, no 7
96	Zak	XI	1	1	1	1	4	12	1	0	0	0	2	12	0.3	0	0	0	0	0	0	0	0	Platon, 88, II, 218, no 1
110	Zak	XLVIII	1	1	1	1	4	12	1	0	0	0	1	11	0.8	0	0	0	0	0	0	0	680	Platon, 88, II, 220, no 15
109	Zak	XLIII	1	1	1	1	4	12	1	0	0	0	3	7	0.5	0	0	0	0	0	0	0	0	Platon, 88, II, 220, no 14
99	Zak	XXVIII	1	1	1	1	4	12	1	0	0	0	1	34	0.3	0	0	0	0	0	0	0	0	Platon, 88, II, 218, no 4
112	Zak	Buil. PD	1	1	2	1	0	12	1	0	0	0	1	4.5	0.5	0	0	0	0	0	0	0	833	Platon, 88, II, 221, no 17
104	Zak	Buil. Ni	1	1	2	1	0	12	1	0	0	0	2	12	0.1	0	0	0	0	0	0	0	0	Platon, 88, II, 219, no 9
97	Zak	XI	1	1	1	1	4	12	1	0	0	0	3	0	0	0	0	0	0	0	0	0	0	Platon, 88, II, 218, no 2
113	Zak	House I	1	1	2	1	0	12	1	0	0	0	1	23	0.5	0	0	50	0	0	0	0	658b	Platon, 88, II, 221, no 18
107	Zak	XLV	1	1	1	1	4	12	1	0	0	0	1	5.5	0.5	0	0	0	0	0	0	0	670	Platon, 88, II, 220, no 12
100	Zak	Buil. Z	1	1	2	1	0	12	1	0	0	0	1	18	0.3	0	0	0	0	0	0	0	0	Platon, 88, II, 218, no 5
106	Zak	XLV	1	1	1	1	4	12	1	0	0	0	1	12	0.5	0	0	0	0	0	0	0	670	Platon, 88, II, 220, no 11
108	Zak	XLIX	1	1	1	1	4	12	1	0	0	0	1	6	0.1	0	0	0	0	0	0	0	0	Platon, 88, II, 220, no 13
98	Zak	XLV	1	1	1	1	4	12	1	0	0	0	1	35	0.7	0	0	0	0	0	0	0	2909	Platon, 88, II, 218, no 3
111	Zak	LVIII	1	1	1	1	4	12	1	0	0	0	1	13	0.5	0	0	0	0	0	0	0	2912	Platon, 88, II, 220, no 16
101	Zak	XLV	1	1	1	1	4	12	1	0	0	0	2	8.5	0.6	0	0	0	0	0	0	0	0	Platon, 88, II, 219, no 6
105	Zak	XLVII	1	1	1	1	4	12	1	0	0	0	1	7	0.1	0	0	0	0	0	0	0	650	Platon, 88, II, 219, no10
1706	Cha	0	4	1	3	2	6	12	1	0	0	0	1	0	0	0	0	0	97	0.8	0.9	0	ChaM	Stos-Gale et.al, 00, 213, Tabl.2; 214, Tabl 3, no 6
1248	Gyp	Tomb IV	2	2	6	2	6	12	1	0	0	0	1	1.1	0	0	0	0	0	0	0	0	0	Hood et al, 58-59, 246, no IV.3
562	Kla	Tomb 2	4	1	6	2	6	12	1	0	0	0	1	10	0.2	0	0	0	0	0	0	0	0	Evely, 93, 90, no 99
537	Kno	UM P	2	2	2	2	5	12	1	0	0	0	1	6.5	0.4	0	0	0	0	0	0	0	0	Evely, 93, 92, no 97
530	Kno	UM H	2	2	2	2	5	12	1	0	0	0	2	3	0.1	0	0	0	0	0	0	0	0	Evely, 93, 90, no 69
529	Kno	UM H	2	2	2	2	5	12	1	0	0	0	2	1	0.1	0	0	0	0	0	0	0	0	Evely, 93, 90, no 68
536	Kno	UM H	2	2	2	2	5	12	1	0	0	0	1	6.8	0.5	0	0	0	97	3	0	0	0	Evely, 93, 92, no 96
531	Kno	UM H	2	2	2	2	5	12	1	0	0	0	2	9	0.1	0	0	0	0	0	0	0	0	Evely, 93, 90, no 70
548	Kno	UM P	2	2	2	2	5	12	1	0	0	0	2	14	0.9	0	0	35	90	9	0	1	0	Evely, 93, 92, no 133
533	Kno	UM M	2	2	2	2	5	12	1	0	0	0	1	3.8	0.2	0	0	0	0	0	0	0	0	Evely, 93, 90, no 72
528	Kno	UM M	2	2	2	2	5	12	1	0	0	0	2	2.3	0.4	0	0	0	0	0	0	0	0	Evely, 93, 90, no 67
527	Kno	UM H	2	2	2	2	5	12	1	0	0	0	2	1.4	0.2	0	0	0	0	0	0	0	0	Evely, 93, 90, no 66
532	Kno	UM M	2	2	2	2	5	12	1	0	0	0	2	5.1	0.2	0	0	0	0	0	0	0	0	Evely, 93, 90, no 71
519	Kno	UM M	2	2	2	2	5	12	1	0	0	0	1	5	0.3	0	0	0	91	9	0	0	0	Evely, 93, 88, no 37
1284	NeH	Tomb V	2	2	6	2	6	12	1	0	0	0	1	3	0	0	0	0	0	0	0	0	0	Hood / d.Jong, 52, 277, no 9
1264	Gyp	Tomb XI	2	2	6	3	8	12	1	0	0	0	1	1.2	0	0	0	0	0	0	0	0	0	Hood et al, 58-59, 251, no XI.4

Appendix V.2 Bronze objects. Sorted by object type / period / site. Total 1857

No	Site	Co	G	L	S	Pe	Da	Ot	C	TM	TK	TP	Cn	Le	Wi	He	Rd	We	Cu	Sn	As	Pb	Mu	Ref
561	Kla	Tomb 4	4	1	6	3	8	12	1	0	0	0	1	20	0.5	0	0	0	0	0	0	0	ChaM	Evely, 93, 90, no 49
549	Kno	Roy. Road	22	2	2	3	7	12	1	0	0	0	1	0	0	0	0	0	0	0	0	0	0	Evely, 93, 92, no 134
550	Kno	Roy. Road	2	2	2	3	0	12	1	0	0	0	1	0	0	0	0	0	0	0	0	0	0	Evely, 93, 92, no 135
1600	Kom	Hou wine P.	3	1	3	0	0	13	1	0	0	0	1	10	0.5	0	0	0	0	0	0	0	0	Blitzer, 95, 516, no M 160
1538	Psy	0	1	2	7	0	0	13	1	0	0	0	1	12	0	0	0	0	0	0	0	0	AshM	Boardman, 61, 33, 34, no M 147
1670	Gou	0	1	1	3	1	4	13	1	0	0	0	1	0	0	0	0	0	0	0	0	0	0	Boyd Hawes, 08, 34, no 38n
1671	Gou	0	1	1	3	1	4	13	1	0	0	0	1	0	0	0	0	0	0	0	0	0	0	Boyd Hawes, 08, 34, no 38n
1673	Gou	Hill House	1	1	3	1	4	13	1	0	0	0	1	15	0	0	0	0	0	0	0	0	0	Boyd Hawes, 08, 34, no 39n
1675	Gou	Hill House	1	1	3	1	4	13	1	0	0	0	1	15	0	0	0	0	0	0	0	0	0	Boyd Hawes, 08, 34, no 39n
1677	Gou	Hill House	1	1	3	1	4	13	1	0	0	0	1	15	0	0	0	0	0	0	0	0	0	Boyd Hawes, 08, 34, no 39n
1674	Gou	Hill House	1	1	3	1	4	13	1	0	0	0	1	15	0	0	0	0	0	0	0	0	0	Boyd Hawes, 08, 34, no 39n
1672	Gou	Hill House	1	1	3	1	4	13	1	0	0	0	1	15	0	0	0	0	0	0	0	0	622	Boyd Hawes, 08, 34, no 39
1668	Gou	Hill House	1	1	3	1	4	13	1	0	0	0	1	15	0	0	0	0	0	0	0	0	622	Boyd Hawes, 08, 34, no 38
1676	Gou	Hill House	1	1	3	1	4	13	1	0	0	0	1	15	0	0	0	0	0	0	0	0	0	Boyd Hawes, 08, 34, no 39n
1669	Gou	0	1	1	3	1	4	13	1	0	0	0	1	0	0	0	0	0	0	0	0	0	0	Boyd Hawes, 08, 34, no 38n
1274	Kcp	0	2	2	6	1	6	13	1	0	0	2	2	9	0	0	0	0	0	0	0	0	0	Hutchinson, 56, 78, 79, no 14
1273	Kcp	0	2	2	6	1	6	13	1	0	0	3	3	5.6	0	0	0	0	0	0	0	0	0	Hutchinson, 56, 78, 79, no 13
1272	Kcp	0	2	2	6	1	6	13	1	0	0	2	2	13	0	0	0	0	0	0	0	0	0	Hutchinson, 56, 78, 79, no 12
1728	Moc	House C2	1	1	3	1	4	13	1	0	0	0	1	10	0	0	0	0	0	0	0	0	0	Soles / Davaras, 94, 399, no CA 16
1727	Moc	House C2	1	1	3	1	4	13	1	0	0	0	1	11	0	0	0	0	0	0	0	0	0	Soles / Davaras, 94, 399, no CA 15
1759	Moc	House C7	1	1	3	1	4	13	1	0	0	0	1	0	0	0	0	0	0	0	0	0	0	Soles / Davaras, 96, 201, no CA 81
1734	Moc	House A2	1	1	3	1	4	13	1	0	0	0	1	0	0	0	0	0	0	0	0	0	0	Soles / Davaras, 94, 419, no CA 97
1508	NiC	Room 7	2	1	4	1	0	13	1	0	0	0	1	0	0	0	0	0	0	0	0	0	0	Georgiou, 79, 43, no 6
1388	Pak	House N	1	1	6	1	4	13	1	0	0	2	2	5.1	0	0	0	0	0	0	0	0	0	Sackett / Popham, 63, 300, no 10
1376	Pak	Kour.	1	1	3	1	4	13	1	0	0	0	1	20	0	0	0	0	0	0	0	0	0	Dawkins / Tod, 02-03, 332
1389	Pak	House N	1	1	6	1	4	13	1	0	0	2	2	5.5	0	0	0	0	0	0	0	0	0	Sackett / Popham, 63, 300, no 11
118	Zak	XXXV	1	1	1	1	4	13	1	0	0	2	2	11	0.4	0	0	0	0	0	0	0	2916	Platon, 88, II, 222, no 1
1598	Kom	Cliffside H	3	1	3	2	6	13	1	0	0	0	1	0	0	0	0	0	0	0	0	0	0	Blitzer, 95, 515, no M 141
1209	Mas	Tomb V	2	2	6	2	6	13	2	0	0	0	1	7.8	0	0	0	0	0	0	0	0	0	Forsdyke, 26-27, 257, no V.3
1221	Mas	Tomb VIIA	2	2	6	2	6	13	1	0	0	2	2	13	0	0	0	0	0	0	0	0	0	Forsdyke, 26-27, 262, no VIIA.7
1126	ZaP	Tomb 32	2	2	6	2	6	13	1	0	0	0	3	0	0	0	0	0	0	0	0	0	0	Evans, 06, 49
1151	ZaP	Tomb 72	2	2	6	2	6	13	2	0	0	3	3	0	0	0	0	0	0	0	0	0	0	Evans, 06, 76, no 72c
1710	Cha	0	4	1	3	3	8	13	2	0	0	0	1	0	0	0	0	0	98	0.8	0.6	0	ChaM	Stos-Gale et.al. 00, 213,Tabl.2; 214,Tabl 3, no 24
1851	EkP	Tomb A	2	2	6	3	0	13	1	0	0	0	1	0	0	0	0	0	0	0	0	0	0	Kanta, 80, 61
1265	Gyp	Tomb XIII	2	2	6	3	8	13	1	0	0	2	2	11	0	0	0	0	0	0	0	0	0	Hood et al, 58-59, 252, no XIII.1
1257	Gyp	Tomb VII	2	2	6	3	8	13	2	0	0	1	1	20	0	0	0	0	0	0	0	0	0	Hood et al, 58-59, 249, no VII.16
1255	Gyp	Tomb VII	2	2	6	3	8	13	2	0	0	2	3	11	0	0	0	0	0	0	0	0	0	Hood et al, 58-59, 249, no VII.14
1256	Gyp	Tomb VII	2	2	6	3	8	13	2	0	0	0	3	7.5	0	0	0	0	0	0	0	0	0	Hood et al, 58-59, 249, no VII.15
1254	Gyp	Tomb VII	2	2	6	3	8	13	2	0	0	1	1	26	0	0	0	0	0	0	0	0	0	Hood et al, 58-59, 249, no VII.13
1423	Kas	0	2	2	6	3	7	13	1	0	0	0	1	0	0	0	0	0	0	0	0	0	0	Kanta, 80, 78

Appendix V.2 Bronze objects. Sorted by object type / period / site. Total 1857

Appendix V.2 / 20

No	Site	Co	G	L	S	Pe	Da	Ot	C	TM	TK	TP	Cn	Le	Wi	He	Rd	We	Cu	Sn	As	Pb	Mu	Ref
1599	Kom	0	3	1	3	3	7	13	1	0	0	0	1	10	0.4	0	0	0	0	0	0	0	0	Blitzer, 95, 517, no M 171
1841	Moc	Tomb 10	1	1	6	3	8	13	2	0	0	0	0	0	0	0	0	0	0	0	0	0	0	Soles / Davaras, 96, 212
1764	Moc	Tomb 10	1	1	6	3	0	13	2	0	0	0	0	0	0	0	0	0	0	0	0	0	0	Soles / Davaras, 96, 212, no CA 93
1438	Paa	Pithos Bur	1	1	6	3	7	13	1	0	0	0	1	0	0	0	0	0	0	0	0	0	0	Kanta, 80, 143
1667	Gou	Hill House	1	1	3	1	4	14	1	0	0	0	1	10	0	0	0	0	0	0	0	0	976	Boyd Hawes, 08, 34, no 25
1596	Kom	0	3	1	3	1	0	14	1	0	0	0	3	3.3	0	0	0	0	0	0	0	0	0	Blitzer, 95, 513, no M 100
1385	Pak	House N	1	1	6	1	4	14	1	0	0	0	0	3.2	0	0	0	0	0	0	0	0	0	Sackett / Popham, 63, 300, no 7
1386	Pak	House N	1	1	6	1	4	14	1	0	0	0	0	8.7	0	0	0	0	0	0	0	0	0	Sackett / Popham, 63, 300, no 8
1387	Pak	House N	1	1	6	1	4	14	1	0	0	0	1	7.8	0	0	0	0	0	0	0	0	0	Sackett / Popham, 63, 300, no 9
1821	Tyl	House A	4	2	4	1	4	14	1	0	0	0	1	0	0	0	0	0	0	0	0	0	0	Driessen / Macdonald, 97, 129
1820	Tyl	House A	4	2	4	1	4	14	1	0	0	0	1	0	0	0	0	0	0	0	0	0	0	Driessen / Macdonald, 97, 129
1819	Tyl	House A	4	2	4	1	4	14	1	0	0	0	1	0	0	0	0	0	0	0	0	0	0	Driessen / Macdonald, 97, 129
115	Zak	S Wing	1	1	1	1	4	14	1	0	0	0	1	7.4	0.5	0	0	0	0	0	0	0	0	Platon, 88, II, 222, no 2
114	Zak	XLV	1	1	1	1	4	14	1	0	0	0	2	14	1.2	0	0	0	0	0	0	0	2908	Platon, 88, II, 221, no 1
116	Zak	S Wing	1	1	1	1	4	14	1	0	0	0	2	9	0.5	0	0	0	0	0	0	0	676	Platon, 88, II, 222, no 3
117	Zak	LXIII	1	1	1	1	3	14	1	0	0	0	1	7	0.5	0	0	0	0	0	0	0	2822	Platon, 88, II, 222, no 4
1287	NeH	Tomb V	2	2	6	2	6	14	2	0	0	0	1	12	0	0	0	0	0	0	0	0	0	Hood / d.Jong, 52, 277, no 12
1288	NeH	Tomb V	2	2	6	2	6	14	1	0	0	0	1	7.3	0	0	0	0	0	0	0	0	0	Hood / d.Jong, 52, 277, no 13
1708	Cha	0	4	1	3	3	7	14	1	0	0	0	3	0	0	0	0	0	91	8.3	0.1	0	ChaM	Stos-Gale et.al., 00, 213, Tabl.2; 214, Tabl 3, no 17
1597	Kom	0	3	1	3	3	8	14	1	0	0	0	1	2.9	0.7	0	0	0	0	0	0	0	0	Blitzer, 95, 513, no M 101
1450	Phk	0	1	2	6	3	7	14	1	0	0	0	1	0	0	0	0	0	0	0	0	0	0	Kanta, 80, 164, 183
1451	Phk	0	1	2	6	3	7	14	1	0	0	0	1	0	0	0	0	0	0	0	0	0	0	Kanta, 80, 164, 183
1720	Arh	Tholos B	2	2	6	0	0	15	1	0	0	0	3	5	1.2	0	0	0	0	0	0	0	HerM	Sakellarakis / Sakellarakis, 97, 602
1594	Kom	0	3	1	3	0	0	15	1	0	0	0	1	5	1.2	0	0	0	0	0	0	0	0	Blitzer, 95, 515, no M 125
1533	Psy	0	1	2	7	0	0	15	1	0	0	0	1	7.2	0	0	0	0	0	0	0	0	AshM	Boardman, 61, 32, no 129
1537	Psy	0	1	2	7	0	0	15	1	0	0	0	3	5	0	0	0	0	0	0	0	0	AshM	Boardman, 61, 32, no 133
1536	Psy	0	1	2	7	0	0	15	1	0	0	0	3	6	0	0	0	0	0	0	0	0	AshM	Boardman, 61, 32, no 132
1534	Psy	0	1	2	7	0	0	15	1	0	0	0	1	6.2	0	0	0	0	0	0	0	0	AshM	Boardman, 61, 32, no 130
1535	Psy	0	1	2	7	0	0	15	1	0	0	0	1	6.3	0	0	0	0	0	0	0	0	AshM	Boardman, 61, 32, no 131
1654	Gou	0	1	1	3	1	4	15	1	0	0	0	1	0	0	0	0	0	0	0	0	0	0	Boyd Hawes, 08, 34, no 32n
1658	Gou	0	1	1	3	1	4	15	1	0	0	0	1	0	0	0	0	0	0	0	0	0	0	Boyd Hawes, 08, 34, no 32n
1651	Gou	0	1	1	3	1	4	15	1	0	0	0	1	0	0	0	0	0	0	0	0	0	0	Boyd Hawes, 08, 34, no 32n
1650	Gou	0	1	1	3	1	4	15	1	0	0	0	1	0	0	0	0	0	0	0	0	0	0	Boyd Hawes, 08, 34, no 32n
1653	Gou	0	1	1	3	1	4	15	1	0	0	0	1	0	0	0	0	0	0	0	0	0	0	Boyd Hawes, 08, 34, no 32n
1652	Gou	0	1	1	3	1	4	15	1	0	0	0	1	0	0	0	0	0	0	0	0	0	0	Boyd Hawes, 08, 34, no 32n
1649	Gou	0	1	1	3	1	4	15	1	0	0	0	1	0	0	0	0	0	0	0	0	0	0	Boyd Hawes, 08, 34, no 32n
1647	Gou	F 29	1	1	3	1	4	15	1	0	0	0	1	8	0	0	0	0	0	0	0	0	941	Boyd Hawes, 08, 34, no 32
1648	Gou	0	1	1	3	1	4	15	1	0	0	0	1	0	0	0	0	0	0	0	0	0	0	Boyd Hawes, 08, 34, no 32n
1268	Kep	0	2	2	6	1	6	15	1	0	0	0	3	0	0	0	0	0	0	0	0	0	0	Hutchinson, 56, 78, 79, no 8
1267	Kep	0	2	2	6	1	6	15	1	0	0	0	1	5.3	0	0	0	0	0	0	0	0	0	Hutchinson, 56, 78, 79, no 7

Appendix V.2   Bronze objects. Sorted by object type / period / site. Total 1857

No	Site	Co	G	L	S	Pe	Da	Ot	C	TM	TK	TP	Cn	Le	Wi	He	Rd	We	Cu	Sn	As	Pb	Mu	Ref
1595	Kom	0	3	1	3	1	3	15	1	0	0	0	1	6.4	1.3	0	0	0	0	0	0	0	0	Blitzer, 95, 515, no M 126
1593	Kom	0	3	1	3	1	4	15	1	0	0	0	2	6.8	1.8	0	0	0	0	0	0	0	0	Blitzer, 95, 512, no M 82
1496	Mal	Quart D	2	2	2	1	0	15	1	0	0	0	1	6.3	0	0	0	0	0	0	0	0	2262	Georgiou ,79, 29, no 6
1751	Moc	House C3	1	1	3	1	4	15	1	0	0	0	1	0	0	0	0	0	0	0	0	0	0	Soles / Davaras, 96, 196, no CA 60
1733	Moc	House A2	1	1	3	1	4	15	1	0	0	0	1	0	0	0	0	0	0	0	0	0	0	Soles / Davaras, 94, 419, no CA 38
1384	Pak	House N	1	1	6	1	0	15	1	0	0	0	3	6.1	0	0	0	0	0	0	0	0	0	Sackett / Popham, 63, 300, no 6
129	Zak	House B	1	1	2	1	0	15	1	0	0	0	3	4	1.8	0	0	0	0	0	0	0	0	Platon, 88, II, 224, no 3
131	Zak	LXVIII	1	1	1	1	4	15	1	0	0	0	1	6.8	1.8	0	0	0	0	0	0	0	0	Platon, 88, II, 225, no 5
128	Zak	XXVIII	1	1	1	1	4	15	1	0	0	0	3	4.8	2.6	0	0	0	0	0	0	0	0	Platon, 88, II, 224, no 2
127	Zak	XXVIII	1	1	1	1	4	15	1	0	0	0	3	7.7	2.4	0	0	0	0	0	0	0	0	Platon, 88, II, 224, no 1
130	Zak	LXVIII	1	1	1	1	4	15	1	0	0	0	3	6.4	2.2	0	0	0	0	0	0	0	0	Platon, 88, II, 224, no 4
1719	Arh	Tholos A	2	2	6	2	6	15	1	0	0	0	1	0	0	0	0	0	0	0	0	0	HcrM	Sakellarakis / Sakellarakis, 97, 602
1484	Kno	UM L	2	2	2	2	5	15	1	0	0	0	1	7.2	0	0	0	0	99	1	0	0	0	Popham, 84, 51, no L 126; Catling / Catling, 84, 215
1222	Mas	Tomb VIIA	2	2	6	2	6	15	1	0	0	0	1	0	0	0	0	0	0	0	0	0	0	Forsdyke, 26-27, 262, no VIIA.8
1210	Mas	Tomb V	2	2	6	2	6	15	1	0	0	0	1	7.2	0	0	0	0	0	0	0	0	0	Forsdyke, 26-27, 257, no V.4
1241	Mas	Tomb XX	2	2	6	2	6	15	1	0	0	0	3	0	0	0	0	0	0	0	0	0	0	Forsdyke, 26-27, 283, no XX.2
1202	Mas	Tomb III	2	2	6	2	6	15	1	0	0	0	2	13	0	0	0	0	0	0	0	0	0	Forsdyke, 26-27, 252, no III.6
1232	Mas	Tom XVIIB	2	2	6	2	6	15	1	0	0	0	2	0	0	0	0	0	0	0	0	0	0	Forsdyke, 26-27, 279, no XVIIB.2
1280	NeH	Tomb I	2	2	6	2	6	15	1	0	0	0	1	8.2	0	0	0	0	0	0	0	0	0	Hood / d.Jong, 52, 265, no 12
1145	ZaP	Tomb 68	2	2	6	2	6	15	1	0	0	0	1	0	0	0	0	0	0	0	0	0	0	Evans, 06, 75, no 68d
1394	Gaz	0	2	1	6	3	8	15	1	0	0	0	1	0	0	0	0	0	0	0	0	0	0	Kanta, 80, 21
1771	Moc	Tomb 15	1	1	6	3	0	15	1	0	0	0	1	0	0	0	0	0	0	0	0	0	0	Soles / Davaras, 96, 222, no CA 114
1407	Stm	Tomb C	2	2	6	3	7	15	1	0	0	0	1	0	0	0	0	0	0	0	0	0	0	Kanta, 80, 53
579	HaT	0	3	2	4	1	0	16	1	0	0	0	1	50	2.5	0	0	0	0	0	0	0	1553	Evely, 93, 101, no 23
44	Zak	XXIV	1	1	1	1	4	16	1	0	0	0	3	12	4.7	0	0	1700	0	0	0	0	2621	Platon, 88, II, 203
51	Zak	House A	1	1	2	1	0	17	1	0	0	0	1	12	3	0	0	310	0	0	0	0	655b	Platon, 88, II, 206, no 6
48	Zak	XXVIII	1	1	1	1	4	17	1	0	0	0	1	17	2.7	0	0	560	0	0	0	0	2625	Platon, 88, II, 205, no 3
53	Zak	House A	1	1	2	1	4	17	1	0	0	0	1	13	2.8	0	0	250	0	0	0	0	655d	Platon, 88, II, 206, no 8
52	Zak	House A	1	1	2	1	0	17	1	0	0	0	1	11	3	0	0	200	0	0	0	0	655c	Platon, 88, II, 206, no 7
50	Zak	House A	1	1	2	1	0	17	1	0	0	0	1	13	3	0	0	370	0	0	0	0	655a	Platon, 88, II, 205, no 5
49	Zak	XXIV	1	1	1	1	4	17	1	0	0	0	1	14	3.1	0	0	460	0	0	0	0	2622	Platon, 88, II, 205, no 4
47	Zak	XXVIII	1	1	1	1	4	17	1	0	0	0	1	21	3.4	0	0	480	0	0	0	0	2624	Platon, 88, II, 205, no 2
46	Zak	XXVIII	1	1	1	1	4	17	1	0	0	0	1	21	3.5	0	0	400	0	0	0	0	2623	Platon, 88, II, 204, no 1
1293	0	0	0	0	0	0	0	18	1	0	0	0	2	20	0	0	0	0	0	0	0	0	AshM	Catling, 68, 95, no 11
1829	HaT	Room 52	3	2	4	1	4	18	1	0	0	0	1	24	0	0	0	0	0	0	0	0	0	Halbherr et.al, 80, 90
60	Zak	Buil. PD	1	1	2	1	0	18	1	0	0	0	1	15	4.7	0	0	0	0	0	0	0	3064	Platon, 88, II, 209, no 2
59	Zak	XV	1	1	1	1	4	18	1	0	0	0	1	17	4.4	0	0	60	0	0	0	0	2597	Platon, 88, II, 209, no 1
61	Zak	Buil. PD	1	1	2	1	0	18	1	0	0	0	1	12	3	0	0	0	0	0	0	0	3065	Platon, 88, II, 210, no 3
1243	Gyp	Tomb I	2	2	6	2	6	18	1	0	0	0	2	20	0	0	0	0	0	0	0	0	0	Hood et al, 58-59, 245, no I.6
1245	Gyp	Tomb I	2	2	6	2	6	18	1	0	0	0	1	19	0	0	0	0	0	0	0	0	0	Hood et al, 58-59, 245, no I.8

Appendix V.2 Bronze objects. Sorted by object type / period / site. Total 1857

No	Site	Co	G	L	S	Pe	Da	Ot	C	TM	TK	TP	Cn	Le	Wi	He	Rd	We	Cu	Sn	As	Pb	Mu	Ref
1250	Gyp	Tomb V	2	2	6	2	6	18	1	0	0	0	1	17	0	0	0	0	0	0	0	0	0	Hood et al, 58-59, 246, no V.1
1279	Hal	Tomb 1	2	2	6	2	5	18	1	0	0	0	2	12	0	0	0	0	0	0	0	0	0	Hood / d.Jong, 52, 262, no 5
1277	Hal	Tomb 2	2	2	6	2	5	18	1	0	0	0	2	19	0	0	0	0	0	0	0	0	0	Hood, 56, 98, no 15
1434	HaT	Tom o Sar	3	2	6	2	6	18	1	0	0	0	1	0	0	0	0	0	0	0	0	0	0	Kanta, 80, 104
1435	HaT	Tom o Sar	3	2	6	2	6	18	1	0	0	0	1	0	0	0	0	0	0	0	0	0	0	Kanta, 80, 104
1179	Iso	Tomb 3	2	2	6	2	6	18	1	0	0	0	1	13	0	0	0	0	0	0	0	0	0	Evans, 14, 15, no 3d
1172	Iso	Tomb 2	2	2	6	2	5	18	1	0	0	0	1	23	0	0	0	0	0	0	0	0	0	Evans, 14, 58, no 2c
1485	Kno	UM L	2	2	2	2	5	18	1	0	0	0	1	18	0	0	0	0	99	1	0	0	0	Popham, 84, 51, no L 128; Catling / Catling, 84, 217
1201	Mas	Tomb III	2	2	6	2	6	18	1	0	0	0	2	0	0	0	0	0	0	0	0	0	0	Forsdyke, 26-27, 252, no III.5
1200	Mas	Tomb III	2	2	6	2	6	18	1	0	0	0	2	0	0	0	0	0	0	0	0	0	0	Forsdyke, 26-27, 252, no III.5
1281	NeH	Tomb III	2	2	6	2	6	18	1	0	0	0	1	17	0	0	0	0	0	0	0	0	0	Hood / d.Jong, 52, 271, no 15
1195	Sel	Tomb 3	2	2	6	2	6	18	1	0	0	0	2	5	0	0	0	0	0	0	0	0	0	Catling / Catling, 74, 240, no 5
1194	Sel	Tomb 3	2	2	6	2	6	18	1	0	0	0	2	7	0	0	0	0	0	0	0	0	0	Catling / Catling, 74, 240, no 4
1197	Sel	Tomb 3	2	2	6	2	6	18	1	0	0	0	3	0	0	0	0	0	0	0	0	0	0	Catling / Catling, 74, 240, no 7
1183	Sel	Tomb 4/I	2	2	6	2	6	18	1	0	0	0	2	15	0	0	0	0	0	0	0	0	0	Catling / Catling, 74, 229, no 7
1196	Sel	Tomb 3	2	2	6	2	6	18	1	0	0	0	2	16	0	0	0	0	0	0	0	0	0	Catling / Catling, 74, 240, no 6
1182	Sel	Tomb 4/I	2	2	6	2	6	18	1	0	0	0	2	16	0	0	0	0	0	0	0	0	0	Catling / Catling, 74, 229, no 6
1123	ZaP	Tomb 21	2	2	6	2	6	18	1	0	0	0	1	17	0	0	0	0	0	0	0	0	0	Evans, 06, 47, no 21b
1127	ZaP	Tomb 33	2	2	6	2	6	18	1	0	0	0	1	0	0	0	0	0	0	0	0	0	0	Evans, 06, 50
1124	ZaP	Tomb 21	2	2	6	2	6	18	1	0	0	0	1	17	0	0	0	0	0	0	0	0	0	Evans, 06, 47, no 21c
1141	ZaP	Tomb 64	2	2	6	2	6	18	1	0	0	0	1	19	0	0	0	0	0	0	0	0	0	Evans, 06, 69, no 64c
1105	ZaP	Tomb 4	2	2	6	2	6	18	1	0	0	0	1	20	0	0	0	0	0	0	0	0	AshM	Evans, 06, 23, no 4a; Baboula / Northover, 99, 149
1156	ZaP	Tomb 78	2	2	6	2	6	18	1	0	0	0	3	0	0	0	0	0	0	0	0	0	0	Evans, 06, 79
1136	ZaP	Tomb 51	2	2	6	2	6	18	1	0	0	0	1	18	0	0	0	0	0	0	0	0	0	Evans, 06, 64, no 51e
1113	ZaP	Tomb 10	2	2	6	2	6	18	1	0	0	0	1	20	0	0	0	0	0	0	0	0	0	Evans, 06, 31, no 10a
1131	ZaP	Tomb 43	2	2	6	2	6	18	1	0	0	0	1	19	0	0	0	0	0	0	0	0	0	Evans, 06, 62, no 43b
1165	ZaP	Tomb 98	2	2	6	2	6	18	1	0	0	0	1	0	0	0	0	0	0	0	0	0	0	Evans, 06, 87, no 98d
1118	ZaP	Tomb 14	2	2	6	2	6	18	1	0	0	0	1	0	0	0	0	0	0	0	0	0	0	Evans, 06, 43, no 14u
1152	ZaP	Tomb 75	2	2	6	2	6	18	1	0	0	0	1	0	0	0	0	0	0	0	0	0	0	Evans, 06, 77, no 75c
1130	ZaP	Tomb 42	2	2	6	2	6	18	1	0	0	0	1	23	0	0	0	0	0	0	0	0	0	Evans, 06, 60, no 42b
1129	ZaP	Tomb 42	2	2	6	2	6	18	1	0	0	0	1	23	0	0	0	0	0	0	0	0	0	Evans, 06, 60, no 42b
1122	ZaP	Tomb 18	2	2	6	2	6	18	1	0	0	0	1	0	0	0	0	0	0	0	0	0	0	Evans, 06, 46
1117	ZaP	Tomb 14	2	2	6	2	6	18	1	0	0	0	1	17	0	0	0	0	0	0	0	0	0	Evans, 06, 43, no 14t
1125	ZaP	Tomb 31	2	2	6	2	6	18	1	0	0	0	1	0	0	0	0	0	0	0	0	0	0	Evans, 06, 49
1106	ZaP	Tomb 4	2	2	6	2	6	18	1	0	0	0	1	0	0	0	0	0	0	0	0	0	1120	Evans, 06, 23, no 4c; Sandars, 55, 189
1144	ZaP	Tomb 68	2	2	6	2	6	18	1	0	0	0	1	16	0	0	0	0	0	0	0	0	0	Evans, 06, 74, no 68a
1121	ZaP	Tomb 18	2	2	6	2	6	18	1	0	0	0	1	0	0	0	0	0	0	0	0	0	0	Evans, 06, 46
1104	ZaP	Tomb 3	2	2	6	2	6	18	1	0	0	0	1	19	0	0	0	0	0	0	0	0	0	Evans, 06, 23, no 3b
1716	Arh	Tholos D	2	2	6	3	7	18	1	0	0	0	2	0	0	0	0	0	0	0	0	0	HerM	Sakellarakis / Sakellarakis, 97, 600
1396	Art	0	2	2	6	3	7	18	1	0	0	0	1	0	0	0	0	0	0	0	0	0	0	Kanta, 80, 45

Appendix V.2 Bronze objects. Sorted by object type / period / site. Total 1857

No	Site	Co	G	L	S	Pe	Da	Ot	C	TM	TK	TP	Cn	Le	Wi	He	Rd	We	Cu	Sn	As	Pb	Mu	Ref
1417	EkP	Tomb A	2	2	6	3	0	18	1	0	0	0	1	0	0	0	0	0	0	0	0	0	0	Kanta, 80, 61
1358	Gal	0	3	2	6	3	7	18	1	0	0	0	1	0	0	0	0	0	0	0	0	0	0	Driessen / Macdonald 84, 71, no Eii 4
1397	Gor	Tomb 2	2	2	6	3	8	18	1	0	0	0	1	0	0	0	0	0	0	0	0	0	0	Kanta, 80, 48
1398	Gor	Tomb 2	2	2	6	3	8	18	1	0	0	0	1	0	0	0	0	0	0	0	0	0	0	Kanta, 80, 48
1263	Gyp	Tomb X	2	2	6	3	8	18	1	0	0	0	1	18	0	0	0	0	0	0	0	0	0	Hood et al, 58-59, 250, no X.4
1253	Gyp	Tomb VI	2	2	6	3	8	18	1	0	0	0	3	3	0	0	0	0	0	0	0	0	0	Hood et al, 58-59, 246, no VI.6
1436	Khf	1961 Tom	3	2	6	3	0	18	1	0	0	0	1	0	0	0	0	0	0	0	0	0	0	Kanta, 80, 105
1422	Kra	0	2	2	5	3	0	18	1	0	0	0	1	0	0	0	0	0	0	0	0	0	0	Kanta, 80, 77
1382	Pak	Bechive T	1	1	6	3	7	18	1	0	0	0	2	13	0	0	0	0	0	0	0	0	0	Bosanquet, 01-02, 304; Sandars, 63, 148
1449	Phk	0	1	2	6	3	7	18	1	0	0	0	1	0	0	0	0	0	0	0	0	0	0	Kanta, 80, 164, 183
1448	Phk	0	1	2	6	3	7	18	1	0	0	0	1	0	0	0	0	0	0	0	0	0	0	Kanta, 80, 164, 183
1447	Phk	0	1	2	6	3	7	18	1	0	0	0	1	0	0	0	0	0	0	0	0	0	0	Kanta, 80, 164, 183
1568	Psy	0	1	2	7	3	0	18	1	0	0	0	2	17	0	0	0	0	0	0	0	0	AshM	Boardman, 61, 51 ,no 223
1566	Psy	0	1	2	7	3	0	18	1	0	0	0	3	6.9	0	0	0	0	88,8	10.9	0.3	0.1	AshM	Boardman, 61, 51, no 221, 160
1567	Psy	0	1	2	7	3	0	18	1	0	0	0	1	12	0	0	0	0	0	0	0	0	AshM	Boardman, 61, 51, no 222
1565	Psy	0	1	2	7	3	0	18	1	0	0	0	1	6.9	0	0	0	0	0	0	0	0	AshM	Boardman, 61, 51, no 220
1413	Stm	Tomb E	2	2	6	3	8	18	1	0	0	0	2	0	0	0	0	0	0	0	0	0	0	Kanta, 80, 54
1405	Stm	Tomb C	2	2	6	3	7	18	1	0	0	0	1	0	0	0	0	0	0	0	0	0	0	Kanta, 80, 53
1406	Stm	Tomb C	2	2	6	3	7	18	1	0	0	0	1	0	0	0	0	0	0	0	0	0	0	Kanta, 80, 53
1411	Stm	Tomb E	2	2	6	3	8	18	1	0	0	0	1	0	0	0	0	0	0	0	0	0	0	Kanta, 80, 54
1403	Stm	Tomb A	2	2	6	3	0	18	1	0	0	0	1	0	0	0	0	0	0	0	0	0	0	Kanta, 80, 53
1412	Stm	Tomb E	2	2	6	3	8	18	1	0	0	0	1	0	0	0	0	0	0	0	0	0	0	Kanta, 80, 54
1426	Tcf	0	2	2	6	3	8	18	1	0	0	0	1	0	0	0	0	0	0	0	0	0	0	Kanta, 80, 80
1607	Gou	0	1	1	3	1	4	19	1	0	0	0	2	9	0	0	0	0	0	0	0	0	557	Boyd Hawes, 08, 34, no 30
1606	Gou	0	1	1	3	1	4	19	1	0	0	0	1	6	0	0	0	0	0	0	0	0	PhiM	Boyd Hawes, 08, 34, no 29
1659	Gou	Hill House	1	1	3	1	4	19	1	0	0	0	1	9	0	0	0	0	0	0	0	0	279	Boyd Hawes, 08, 34, no 31
1605	Gou	0	1	1	3	1	4	19	1	0	0	0	1	9	0	0	0	0	0	0	0	0	557	Boyd Hawes, 08, 34, no 28
79	Zak	W Court	1	1	1	1	4	19	1	0	0	0	2	3.7	1.7	0	0	0	0	0	0	0	0	Platon, 88, II, 214, no 3
74	Zak	Road	1	1	2	1	0	19	1	0	0	0	1	4.7	2.4	0	0	0	0	0	0	0	3098	Platon, 88, II, 213, no 5
77	Zak	House B	1	1	2	1	0	19	1	0	0	0	1	5.5	1.8	0	0	0	0	0	0	0	0	Platon, 88, II, 214, no 1
70	Zak	XXVIII	1	1	1	1	4	19	1	0	0	0	2	13	1.9	0	0	0	0	0	0	0	0	Platon, 88, II, 212, no 1
72	Zak	East Buil.	1	1	2	1	0	19	1	0	0	0	2	6.5	2.2	0	0	0	0	0	0	0	0	Platon, 88, II, 213, no 3
73	Zak	House B	1	1	2	1	0	19	1	0	0	0	2	7.3	0.7	0	0	0	0	0	0	0	0	Platon, 88, II, 213, no 4
76	Zak	XLIV	1	1	1	1	4	19	1	0	0	0	1	6	2.9	0	0	6	0	0	0	0	2910	Platon, 88, II, 214, no 7
71	Zak	XXVIII	1	1	1	1	4	19	1	0	0	0	1	6.5	1.9	0	0	0	0	0	0	0	0	Platon, 88, II, 212, no 2
75	Zak	XLVIII	1	1	1	1	4	19	1	0	0	0	3	6	0	0	0	0	0	0	0	0	0	Platon, 88, II, 213, no 6
78	Zak	XLIV	1	1	1	1	4	19	1	0	0	0	3	6	1.7	0	0	0	0	0	0	0	0	Platon, 88, II, 214, no 2
1186	Sel	Tomb 4/I	2	2	6	2	6	19	1	0	0	0	1	10	0	0	0	0	0	0	0	0	0	Catling / Catling, 74, 229, no 10
1692	Gou	0	1	1	3	1	4	20	1	0	0	0	1	0	0	0	0	0	0	0	0	0	0	Boyd Hawes, 08, 34, no 42n
1683	Gou	0	1	1	3	1	4	20	1	0	0	0	1	0	0	0	0	0	0	0	0	0	0	Boyd Hawes, 08, 34, no 42n

Appendix V.2 Bronze objects. Sorted by object type / period / site. Total 1857

No	Site	Co	G	L	S	Pe	Da	Ot	C	TM	TK	TP	Cn	Le	Wi	He	Rd	We	Cu	Sn	As	Pb	Mu	Ref
1613	Gou	0	1	1	3	1	4	20	1	0	0	0	1	0	0	0	0	0	0	0	0	0	0	Boyd Hawes, 08, 34, no 34n
1679	Gou	G 23	1	1	3	1	4	20	1	0	0	0	1	11	0	0	0	0	0	0	0	0	569	Boyd Hawes, 08, 34, no 18
1691	Gou	0	1	1	3	1	4	20	1	0	0	0	1	0	0	0	0	0	0	0	0	0	0	Boyd Hawes, 08, 34, no 42n
1693	Gou	0	1	1	3	1	4	20	1	0	0	0	1	0	0	0	0	0	0	0	0	0	0	Boyd Hawes, 08, 34 ,no 42n
1682	Gou	0	1	1	3	1	4	20	1	0	0	0	1	17	0	0	0	0	0	0	0	0	623	Boyd Hawes, 08, 34, no 42
1684	Gou	0	1	1	3	1	4	20	1	0	0	0	1	0	0	0	0	0	0	0	0	0	0	Boyd Hawes, 08, 34, no 42n
1608	Gou	Fg	1	1	3	1	4	20	1	0	0	0	1	16	0	0	0	0	0	0	0	0	599	Boyd Hawes, 08, 34, no 33
1687	Gou	0	1	1	3	1	4	20	1	0	0	0	1	0	0	0	0	0	0	0	0	0	0	Boyd Hawes, 08, 34, no 42n
1688	Gou	0	1	1	3	1	4	20	1	0	0	0	1	0	0	0	0	0	0	0	0	0	0	Boyd Hawes, 08, 34, no 42n
1609	Gou	0	1	1	3	1	4	20	1	0	0	0	1	15	0	0	0	0	0	0	0	0	597	Boyd Hawes, 08, 34, no 34
1611	Gou	0	1	1	3	1	4	20	1	0	0	0	1	20	0	0	0	0	0	0	0	0	596	Boyd Hawes, 08, 34, no 36
1610	Gou	Fg	1	1	3	1	4	20	1	0	0	0	1	10	0	0	0	0	0	0	0	0	598	Boyd Hawes, 08, 34, no 35
1690	Gou	0	1	1	3	1	4	20	1	0	0	0	1	0	0	0	0	0	0	0	0	0	0	Boyd Hawes, 08, 34, no 42n
1850	Gou	0	1	1	3	1	4	20	1	0	0	0	1	13	0	0	0	0	0	0	0	0	PhiM	Boyd Hawes, 08, 34, no 41
1612	Gou	C 31	1	1	3	1	4	20	1	0	0	0	1	20	0	0	0	0	0	0	0	0	938	Boyd Hawes, 08, 34, no 37
1678	Gou	G 23	1	1	3	1	4	20	1	0	0	0	1	2.3	0	0	0	0	0	0	0	0	945	Boyd Hawes, 08, 34, no 8
1849	Gou	C 9	1	1	3	1	4	20	1	0	0	0	1	12	0	0	0	0	0	0	0	0	955	Boyd Hawes, 08, 34, no 40
1680	Gou	0	1	1	3	1	4	20	1	0	0	0	1	9.7	0	0	0	0	0	0	0	0	959	Boyd Hawes, 08, 34, no 26
1689	Gou	0	1	1	3	1	4	20	1	0	0	0	1	0	0	0	0	0	0	0	0	0	0	Boyd Hawes, 08, 34, no 42n
1686	Gou	0	1	1	3	1	4	20	1	0	0	0	1	0	0	0	0	0	0	0	0	0	0	Boyd Hawes, 08, 34, no 42n
1685	Gou	0	1	1	3	1	4	20	1	0	0	0	1	0	0	0	0	0	0	0	0	0	0	Boyd Hawes, 08, 34, no 42n
1678	Gou	E 6	1	1	3	1	4	20	1	0	0	0	1	9.2	0	0	0	0	0	0	0	0	0	Boyd Hawes, 08, 34, no 7
1380	Pak	Kour.	1	1	3	1	4	20	1	0	0	0	1	0	0	0	0	0	0	0	0	0	0	Dawkins / Tod, 02–03, 333
1373	Pak	Block Ksi	1	1	3	1	4	20	1	0	0	0	1	0	0	0	0	0	0	0	0	0	0	Driessen / Macdonald, 97, 230
1369	Pak	Block Chi	1	1	3	1	4	20	1	0	0	0	1	0	0	0	0	0	0	0	0	0	0	Driessen / Macdonald, 97, 228
1381	Pak	Kour.	1	1	3	1	4	20	1	0	0	0	1	0	0	0	0	0	0	0	0	0	0	Dawkins / Tod, 02–03, 333
82	Zak	Buil. PD	1	1	2	1	0	20	1	0	0	0	1	9.3	0.9	0	0	0	0	0	0	0	3097	Platon, 88, II, 215, no 1
1601	Kom	0	3	1	3	3	6	20	1	0	0	0	2	20	2.7	0	0	0	0	0	0	0	0	Blitzer, 95, 517, no M 169
1410	Pak	Block Chi	1	1	3	3	7	20	1	0	0	0	1	0	0	0	0	0	0	0	0	0	0	Kanta, 80, 191
783	Kno	0	2	2	2	0	0	21	2	11	0	0	2	0	0	6.5	33	0	0	0	0	0	1082	Matthäus, 80, 10, 210, no 322
797	Mal	Azymo	2	1	2	0	0	21	2	2	0	0	2	0	0	47	44	0	0	0	0	0	2085	Matthäus, 80, 103, no 60, 105
796	Mal	Azymo	2	1	2	0	0	21	2	2	0	0	1	0	0	42	37	0	0	0	0	0	2086	Matthäus, 80, 103, no 59, 105
875	Pha	0	3	2	1	0	0	21	2	2	0	0	3	0	0	0	0	0	0	0	0	0	HerM	Matthäus, 80, 103, no 62
877	Pha	0	3	2	1	0	0	21	2	18	0	0	3	0	0	0	0	0	0	0	0	0	HerM	Matthäus, 80, 284, no 427
876	Pha	0	3	2	2	0	0	21	2	2	0	0	2	0	0	12	27	0	98	0	0	0	HerM	Matthäus, 80, 104, no 63, 105
930	Pik	0	1	2	2	0	0	21	2	11	0	0	2	0	0	7.4	33	0	0	0	0	0	HNM	Matthäus, 80, 207, no 311
821	Zak	Tomb D	1	1	6	0	0	21	2	2	0	0	3	0	0	0	0	0	0	0	0	0	HerM	Matthäus, 80, 112, no 95
935	Cre	0	1	0	0	1	0	21	2	2	0	0	2	0	0	7.5	31	0	0	0	0	0	AshM	Matthäus, 80, 126, no 124, 128
937	Cre	0	1	0	0	1	0	21	2	5	0	0	1	0	0	5.1	27	0	0	0	0	0	AshM	Matthäus, 80, 142, no 167, 143
934	Cre	0	1	0	0	1	0	21	2	2	0	0	2	0	0	28	28	0	0	0	0	0	AshM	Matthäus, 80, 104, no 69, 70

Appendix V.2 Bronze objects. Sorted by object type / period / site. Total 1857

Appendix V.2 / 25

No	Site	Co	G	L	S	Pe	Da	Ot	C	TM	TK	TP	Cn	Le	Wi	He	Rd	We	Cu	Sn	As	Pb	Mu	Ref
936	Cre	0	1	0	0	1	0	21	2	5	0	0	1	0	0	6	26	0	0	0	0	0	AshM	Matthäus, 80, 142, no 166, 143
835	Gou	C 30	1	1	3	1	4	21	2	23	0	0	3	0	0	0	0	0	0	0	0	0	HerM	Matthäus, 80. 320, no 583
833	Gou	F 41	1	1	3	1	4	21	2	4	0	0	3	0	0	0	0	0	0	0	0	0	HerM	Matthäus, 80, 132, no 158
831	Gou	B 6	1	1	3	1	4	21	2	2	0	0	1	0	0	26	26	0	0	0	0	0	605	Matthäus, 80, 102, no 46
832	Gou	F 18	1	1	3	1	4	21	2	2	0	0	3	0	0	0	0	0	0	0	0	0	HerM	Matthäus, 80, 103, no 47
836	Gou	A 43	1	1	3	1	4	21	2	23	0	0	3	0	0	0	0	0	0	0	0	0	HerM	Matthäus, 80, 320, no 584
834	Gou	G 15	1	1	3	1	4	21	2	18	0	0	3	0	0	0	0	0	0	0	0	0	617	Matthäus, 80, 284, no 426
1785	HaT	Ro 31-32	3	2	4	1	4	21	2	1	0	0	2	0	0	0	0	0	0	0	0	0	HerM	Evely, 93, 85, no16; Driessen / Macdonald, 97, 202
1789	HaT	Lightwel 9	3	2	4	1	4	21	2	0	0	0	2	0	0	0	0	0	0	0	0	0	0	Driessen / Macdonald, 97, 202
1787	HaT	Lightwel 9	3	2	4	1	4	21	2	0	0	0	2	0	0	0	0	0	0	0	0	0	0	Driessen / Macdonald, 97, 202
1786	HaT	Lightwel 9	3	2	4	1	4	21	2	0	0	0	2	0	0	0	0	0	0	0	0	0	0	Driessen / Macdonald, 97, 202
1788	HaT	Lightwel 9	3	2	4	1	4	21	2	0	0	0	2	0	0	0	0	0	0	0	0	0	0	Driessen / Macdonald, 97, 202
1791	HaT	Lightwel 9	3	2	4	1	4	21	2	0	0	0	2	0	0	0	0	0	0	0	0	0	0	Driessen / Macdonald, 97, 202
1790	HaT	Lightwel 9	3	2	4	1	4	21	2	0	0	0	2	0	0	0	0	0	0	0	0	0	0	Driessen / Macdonald, 97, 202
1782	HaT	Casa d Lcb	3	2	3	1	4	21	2	2	0	0	1	0	0	52	0	0	0	0	0	0	HerM	Evely, 93, 103, no 48; Driessen / Macdonald, 97, 204
1784	HaT	Room 27	3	2	4	1	4	21	2	1	0	0	1	0	0	6.5	0	0	0	0	0	0	HerM	Evely, 93, 85, no 15; Driessen / Macdonald, 97, 202
1792	HaT	Hall 12	3	2	4	1	4	21	2	0	0	0	2	0	0	0	0	0	0	0	0	0	0	Driessen / Macdonald, 97, 202
761	Kno	NW THo	2	2	2	1	3	21	2	11	0	0	1	0	0	0	33	0	0	0	0	0	HerM	Matthäus, 80, 211, no324; Driessen / Macdonald, 97, 155
778	Kno	SW o SHou	2	2	2	1	3	21	2	11	0	0	1	0	0	0	0	0	0	0	0	0	HerM	Matthäus, 80, 10, 213, no 332
764	Kno	NW House	2	2	2	1	3	21	2	2	0	0	1	0	0	38	38	0	0	0	0	0	HerM	Matthäus, 80, 103, no 55; Driessen / Macdonald, 97, 154
775	Kno	SW o SHou	2	2	2	1	3	21	2	11	0	0	1	0	0	0	30	0	0	0	0	0	HerM	Matthäus, 80, 10, 207, no 310
759	Kno	NW THo	2	2	2	1	3	21	2	9	0	0	1	0	0	35	12	0	0	0	0	0	843	Matthäus, 80, 178, no252; Driessen / Macdonald, 97, 155
767	Kno	SE o SHou	2	2	2	1	0	21	2	2	0	0	1	0	0	52	40	0	0	0	0	0	HerM	Matthäus, 80, 103, no 57; Georgiou, 79, 25
763	Kno	NW THo	2	2	2	1	3	21	2	11	0	0	2	0	0	6.5	32	0	0	0	0	0	845	Matthäus, 80, 211, no 326; Driessen / Macdonald, 97,155
769	Kno	SW o SHou	2	2	2	1	3	21	2	3	0	0	1	0	0	0	30	0	0	0	0	0	HerM	Matthäus, 80, 10, 123, no 106
773	Kno	SW o SHou	2	2	2	1	3	21	2	3	0	0	1	0	0	0	30	0	0	0	0	0	HerM	Matthäus, 80, 10, 123, no 110
768	Kno	SE o SHou	2	2	2	1	0	21	2	8	0	0	1	0	0	61	18	0	0	0	0	0	HerM	Matthäus, 80, 163, no 210; Georgiou, 79, 25
776	Kno	SW o SHou	2	2	2	1	3	21	2	11	0	0	1	0	0	0	30	0	0	0	0	0	HerM	Matthäus, 80, 10, 208, no 314
762	Kno	NW THo	2	2	2	1	3	21	2	11	0	0	1	0	0	8	38	0	0	0	0	0	844	Matthäus, 80, 211, no325; Driessen / Macdonald, 97, 155
760	Kno	NW THo	2	2	2	1	3	21	2	11	0	0	2	0	0	0	30	0	0	0	0	0	HerM	Matthäus, 80, 207, no309; Driessen / Macdonald, 97, 155
771	Kno	SW o SHou	2	2	2	1	3	21	2	3	0	0	1	0	0	0	30	0	0	0	0	0	HerM	Matthäus, 80, 10, 123, no 108
774	Kno	SW o SHou	2	2	2	1	3	21	2	10	0	0	1	0	0	18	14	0	0	0	0	0	2074	Matthäus, 80, 10, 200, no 301
772	Kno	SW o SHou	2	2	2	1	3	21	2	3	0	0	1	0	0	0	30	0	0	0	0	0	HerM	Matthäus, 80, 10, 123, no 109
780	Kno	SW o SHou	2	2	2	1	3	21	2	17	0	0	1	0	0	8.4	18	0	87.2	6.7	0.1	0	2075	Matthäus, 80, 10, 261, no 370; Mangou / Ioannou, 98, 94
779	Kno	SW o SHou	2	2	2	1	3	21	2	11	0	0	1	0	0	0	0	0	0	0	0	0	HerM	Matthäus, 80, 10, 213, no 333
765	Kno	SE o SHou	2	2	2	1	0	21	2	1	0	0	1	0	0	15	30	0	0	0	0	0	HerM	Matthäus, 80, 96, no 32
766	Kno	SE o SHou	2	2	2	1	0	21	2	2	0	0	1	0	0	60	61	0	0	0	0	0	HerM	Matthäus, 80, 103, no 56; Georgiou, 79, 25
770	Kno	SW o SHou	2	2	2	1	3	21	2	2	0	0	1	0	0	0	30	0	0	0	0	0	HerM	Matthäus, 80, 10, 123, no 107
777	Kno	SW o SHou	2	2	2	1	3	21	2	11	0	0	1	0	0	0	30	0	0	0	0	0	HerM	Matthäus, 80, 10, 208, no 315
927	Kou	0	0	4	0	1	0	21	2	8	0	0	2	0	0	49	0	0	0	0	0	0	RetM	Matthäus, 80, 163, no 213

Appendix V.2   Bronze objects. Sorted by object type / period / site. Total 1857

No	Site	Co	G	L	S	Pe	Da	Ot	C	TM	TK	TP	Cn	Le	Wi	He	Rd	We	Cu	Sn	As	Pb	Mu	Ref
793	Mal	Gramka	2	1	2	1	2	21	2	23	0	0	3	0	0	0	0	0	0	0	0	0	HerM	Matthäus, 80, 11, 311, no 487
784	Mal	Gramka	2	1	2	1	2	21	2	2	0	0	3	0	0	17	30	0	0	0	0	0	HerM	Matthäus, 80, 11, 103, no 61
788	Mal	Gramka	2	1	2	1	2	21	2	8	0	0	1	0	0	40	11	0	0	0	0	0	HerM	Matthäus, 80, 11, 163, no 211
798	Mal	Hou Db / 8	2	1	2	1	3	21	2	1	0	0	0	0	0	0	40	0	0	0	0	0	HerM	Matthäus, 80, 83, no 2
792	Mal	Gramka	2	1	2	1	2	21	2	12	0	0	3	0	0	0	0	0	0	0	0	0	HerM	Matthäus, 80, 11, 225, no 347
791	Mal	Gramka	2	1	2	1	2	21	2	11	0	0	2	0	0	7.7	39	0	0	0	0	0	2193	Matthäus, 80, 11, 211, no 327
785	Mal	Gramka	2	1	2	1	2	21	2	3	0	0	2	0	0	4.9	22	0	0	0	0	0	2194	Matthäus, 80, 11, 123, no 111
794	Mal	Gramka	2	1	2	1	2	21	2	23	0	0	3	0	0	0	0	0	0	0	0	0	HerM	Matthäus, 80, 11, 321, no 590
787	Mal	Gramka	2	1	2	1	2	21	2	3	0	0	1	0	0	3.7	16	0	0	0	0	0	2196	Matthäus, 80, 11, 125, no 118
786	Mal	Gramka	2	1	2	1	2	21	2	3	0	0	2	0	0	5.1	22	0	0	0	0	0	2195	Matthäus, 80, 11, 125, no 117
795	Mal	Quart.L/XV	2	1	2	1	3	21	2	3	0	0	2	0	0	45	0	0	100	0	0	0	HerM	Matthäus, 80, 103, no58, 105
789	Mal	Gramka	2	1	2	1	3	21	2	10	0	0	1	0	0	18	13	0	0	0	0	0	2192	Matthäus, 80, 12, 200, no 302
790	Mal	Gramka	2	1	2	1	2	21	2	11	0	0	3	0	0	0	25	0	0	0	0	0	HerM	Matthäus, 80, 11, 209, no 320
805	Moc	Tomb IX	1	1	6	1	0	21	2	12	0	0	2	0	0	0	0	0	0	0	0	0	HerM	Matthäus, 80, 219, no 339
1739	Moc	Build. B2	1	1	3	1	4	21	2	11	0	0	1	0	0	5	25	0	0	0	0	0	0	Soles / Davaras, 96, 193, no CA 109
806	Moc	Tomb IX	1	1	6	1	0	21	2	12	0	0	2	0	0	0	0	0	0	0	0	0	HerM	Matthäus, 80, 219, no 340
1741	Moc	Build. B2	1	1	3	1	4	21	2	11	0	0	1	0	0	5	30	0	0	0	0	0	0	Soles / Davaras, 96, 193, no CA 112
1757	Moc	House C7	1	1	3	1	4	21	2	0	0	0	3	0	0	0	0	0	0	0	0	0	0	Soles / Davaras, 96, 201, no CA 79
809	Moc	Tomb XII	1	1	6	1	4	21	2	18	0	0	2	0	0	4	19	0	0	0	0	0	HerM	Matthäus, 80, 277, no 415
808	Moc	Tomb VII	1	1	6	1	0	21	2	17	0	0	1	0	0	8.8	22	0	0	0	0	0	HerM	Matthäus, 80, 260, no 369
802	Moc	Bloc B/C	1	1	3	1	4	21	2	11	0	0	1	0	0	0	0	0	0	0	0	0	HerM	Matthäus, 80, 213, no 334
1726	Moc	House C4	1	1	3	1	4	21	2	3	0	0	1	0	0	6.8	31	0	0	0	0	0	0	Soles / Davaras, 94, 400, no CA 25
1742	Moc	House C3	1	1	3	1	4	21	2	0	0	0	3	0	0	0	0	0	0	0	0	0	0	Soles / Davaras, 96, 194, no CA 58
1756	Moc	House C7	1	1	3	1	4	21	2	0	0	0	3	0	0	0	0	0	0	0	0	0	0	Soles / Davaras, 96, 201, no CA 76
1725	Moc	House A2	1	1	3	1	4	21	2	12	0	0	1	0	0	0	0	0	0	0	0	0	0	Soles / Davaras, 94, 415, no CA 19
1740	Moc	Build. B2	1	1	3	1	4	21	2	11	0	0	2	0	0	5	30	0	0	0	0	0	0	Soles / Davaras, 96, 193, no CA 111
799	Moc	Bloc B/C	1	1	3	1	4	21	2	3	0	0	1	0	0	7.3	30	0	100	0	0	0	1581	Matthäus, 80, 123, no 112, 124
1737	Moc	Build. B2	1	1	3	1	4	21	2	11	0	0	1	0	0	5	27	0	0	0	0	0	0	Soles / Davaras, 96, 193, no CA 107
801	Moc	Bloc B/C	1	1	3	1	4	21	2	11	0	0	2	0	0	6.4	35	0	0	0	0	0	1578	Matthäus, 80, 211, no 328, 212
800	Moc	Bloc B/C	1	1	3	1	4	21	2	3	0	0	1	0	0	0	0	0	0	0	0	0	1582	Matthäus, 80, 123, no 113, 124
1736	Moc	Build. B2	1	1	3	1	4	21	2	3	0	0	1	0	0	5	25	0	0	0	0	0	0	Soles / Davaras, 96, 193, no CA 106
804	Moc	Tomb IX	1	1	6	1	0	21	2	12	0	0	1	0	0	3.6	14	0	0	0	0	0	HerM	Matthäus, 80, 219, no 338
1724	Moc	House A2	1	1	3	1	4	21	2	12	0	0	1	0	0	5	0	0	0	0	0	0	0	Soles / Davaras, 94, 415, no CA 18
803	Moc	Bloc B/C	1	1	3	1	4	21	2	11	0	0	1	0	0	0	0	0	0	0	0	0	HerM	Matthäus, 80, 213, no 335
1738	Moc	Build. B2	1	1	3	1	4	21	2	11	0	0	1	0	0	5	25	0	0	0	0	0	0	Soles / Davaras, 96, 193, no CA 108
807	Moc	Tomb XII	1	1	6	1	4	21	2	13	0	0	1	0	0	6.1	12	0	0	0	0	0	0	Matthäus, 80, 238, no 357
923	Pak	Ksi Ro 14	1	1	3	1	4	21	2	8	0	0	0	0	0	55	0	0	0	0	0	0	HerM	Matthäus, 80, 163, no 232, 168
924	Pak	N, Room 9	1	1	3	1	4	21	2	18	0	0	2	0	0	7	11	0	0	0	0	0	HerM	Matthäus, 80, 283, no 424
928	Ret	0	4	4	0	1	0	21	2	8	0	0	2	0	0	48	0	0	0	0	0	0	RetM	Matthäus, 80, 163, no 214
814	Tyl	0	4	2	4	1	0	21	2	13	0	0	3	0	0	0	0	0	0	0	0	0	HerM	Matthäus, 80, 239, no 358

Appendix V.2 Bronze objects. Sorted by object type / period / site. Total 1857

No	Site	Co	G	L	S	Pe	Da	Ot	C	TM	TK	TP	Cn	Le	Wi	He	Rd	We	Cu	Sn	As	Pb	Mu	Ref
810	Tyl	House A	4	2	4	1	4	21	2	1	0	0	0	1	0	0	48	97	24000	100	0	0	0	HerM	Matthäus, 80, 82, no1; Driessen / Macdonald; 97, 129
812	Tyl	House A	4	2	4	1	4	21	2	1	0	0	0	1	0	0	33	71	15000	0	0	0	0	HerM	Matthäus, 80, 83, no 4
811	Tyl	House A	4	2	4	1	4	21	2	1	0	0	0	1	0	0	44	125	52000	0	0	0	0	HerM	Matthäus, 80, 83, no3; Driessen / Macdonald, 97, 129
813	Tyl	House A	4	2	4	1	4	21	2	1	0	0	0	1	0	0	35	84	0	0	0	0	0	HerM	Matthäus, 80, 85, no 14
815	Zak	XII	1	1	1	1	4	21	2	3	0	0	0	3	0	0	0	0	0	0	0	0	0	HerM	Matthäus, 80, 103, no 49
827	Zak	XLVa	1	1	1	1	4	21	2	21	0	0	0	1	0	0	0	43	0	0	0	0	0	HerM	Matthäus, 80, 304, no 466
826	Zak	XXVIII	1	1	1	1	4	21	2	20	0	0	0	1	0	0	0	0	0	0	0	0	0	HerM	Matthäus, 80, 300, no 453
822	Zak	Tomb 1962	1	1	1	1	4	21	2	2	0	0	0	3	0	0	0	0	0	0	0	0	0	HerM	Matthäus, 80, 126, no 123
819	Zak	XLVa	1	1	1	1	4	21	2	2	0	0	0	3	0	0	0	0	0	0	0	0	0	HerM	Matthäus, 80, 103, no 53
820	Zak	XLVIII	1	1	1	1	4	21	2	2	0	0	0	2	0	0	0	0	0	0	0	0	0	HerM	Matthäus, 80, 103, no 54
823	Zak	XLVa	1	1	1	1	4	21	2	9	0	0	0	2	0	0	0	30	0	0	0	0	0	HerM	Matthäus, 80, 181, no 262
825	Zak	XLVb	1	1	1	1	4	21	2	19	0	0	0	3	0	0	0	0	0	0	0	0	0	HerM	Matthäus, 80, 297, no 451
829	Zak	XLVa	1	1	1	1	4	21	2	22	0	0	0	1	0	0	0	0	0	0	0	0	0	HerM	Matthäus, 80, 306, no 471
818	Zak	XLVa	1	1	1	1	4	21	2	2	0	0	0	2	0	0	0	0	0	0	0	0	0	HerM	Matthäus, 80, 103, no 52
816	Zak	XLII	1	1	1	1	4	21	2	2	0	0	0	2	0	0	0	0	0	0	0	0	0	HerM	Matthäus, 80, 103, no 50
830	Zak	House A /P	1	1	2	1	4	21	2	22	0	0	0	1	0	0	0	0	0	0	0	0	0	HerM	Matthäus, 80, 306, no 472
824	Zak	XXV	1	1	1	1	4	21	2	11	0	0	0	3	0	0	0	30	0	0	0	0	0	HerM	Matthäus, 80, 211, no 323
817	Zak	XLIV	1	1	1	1	4	21	2	2	0	0	0	2	0	0	0	0	0	0	0	0	0	HerM	Matthäus, 80, 103, no 51
828	Zak	XXVII	1	1	1	1	4	21	2	22	0	0	0	1	0	0	0	0	0	0	0	0	0	HerM	Matthäus, 80, 306, no 470
838	Arh	Tholos A	2	2	6	2	6	21	2	2	0	0	0	1	0	0	30	0	0	0	0	0	0	HerM	Matthäus, 80, 108, no 71
846	Arh	Tholos A	2	2	6	2	6	21	2	20	0	0	0	1	0	0	0	0	0	0	0	0	0	HerM	Matthäus, 80, 302, no 461
842	Arh	Tholos A	2	2	6	2	6	21	2	9	0	0	0	1	0	0	20	0	0	0	0	0	0	HerM	Matthäus, 80, 194, no 292
840	Arh	Tholos A	2	2	6	2	6	21	2	8	0	0	0	1	0	0	0	14	0	0	0	0	0	HerM	Matthäus, 80, 172, no 237
843	Arh	Tholos A	2	2	6	2	6	21	2	12	0	0	0	1	0	0	7	0	0	0	0	0	0	HerM	Matthäus, 80, 235, no 353
845	Arh	Tholos A	2	2	6	2	6	21	2	17	0	0	0	1	0	0	15	0	0	0	0	0	0	HerM	Matthäus, 80, 265, no 395
841	Arh	Tholos A	2	2	6	2	6	21	2	9	0	0	0	1	0	0	22	0	0	0	0	0	0	HerM	Matthäus, 80, 194, no 291
837	Arh	Tholos A	2	2	6	2	6	21	2	1	0	0	0	1	0	0	10	33	0	0	0	0	0	HerM	Matthäus, 80, 96, no 35
839	Arh	Tholos A	2	2	6	2	6	21	2	5	0	0	0	1	0	0	0	10	0	0	0	0	0	HerM	Matthäus, 80, 145, no 175
844	Arh	Tholos A	2	2	6	2	6	21	2	17	0	0	0	1	0	0	12	0	0	0	0	0	0	HerM	Matthäus, 80, 262, no 371
860	Cha	S of Just.	4	1	6	2	6	21	2	1	0	0	0	1	0	0	9.8	31	0	0	0	0	0	ChaM	Matthäus, 80, 96, no 30
863	Cha	S of Just	4	1	6	2	6	21	2	2	0	0	0	3	0	0	0	0	0	0	0	0	0	ChaM	Matthäus, 80, 102, no 45
862	Cha	S of Just	4	1	6	2	6	21	2	2	0	0	0	1	0	0	33	30	0	0	0	0	0	ChaM	Matthäus, 80, 102, no 44
865	Cha	S of Just	4	1	6	2	6	21	2	18	0	0	0	1	0	0	5.3	22	0	0	0	0	0	ChaM	Matthäus, 80, 285, no 430
861	Cha	S of Just	4	1	6	2	6	21	2	1	0	0	0	1	0	0	11	36	0	0	0	0	0	ChaM	Matthäus, 80, 96, no 36
864	Cha	S of Just	4	1	6	2	6	21	2	17	0	0	0	1	0	0	15	20	0	0	0	0	0	ChaM	Matthäus, 80, 266, no 403
866	Cha	Site Kapet	4	1	6	2	6	21	2	8	0	0	0	1	0	0	44	14	0	0	0	0	0	ChaM	Matthäus, 80, 172, no 238
932	Kat	Tomb B	2	1	6	2	5	21	2	18	0	0	0	2	0	0	4.5	13	0	0	0	0	0	HerM	Matthäus, 80, 277, no 414
868	Kly	0	3	2	6	2	6	21	2	1	0	0	0	2	0	0	10	35	0	0	0	0	0	HerM	Matthäus, 80, 39, 96, no 31
871	Kly	Tomb 4	3	2	6	2	6	21	2	8	0	0	0	1	0	0	35	13	0	0	0	0	0	HerM	Matthäus, 80, 39, 172, no 240
870	Kly	0	3	2	6	2	6	21	2	3	0	0	0	3	0	0	0	56	0	0	0	0	0	HerM	Matthäus, 80, 39, 130, no 150

Appendix V.2 / 28

Appendix V.2   Bronze objects. Sorted by object type / period / site. Total 1857

No	Site	Co	G	L	S	Pe	Da	Ot	C	TM	TK	TP	Cn	Le	Wi	He	Rd	We	Cu	Sn	As	Pb	Mu	Ref
873	Kly	Tomb 8	3	2	6	2	6	21	2	9	0	0	2	0	0	13	0	0	0	0	0	0	HerM	Matthäus, 80, 39, 197, no 297
874	Kly	Tomb 8	3	2	6	2	6	21	2	17	0	0	2	0	0	10	18	0	0	0	0	0	HerM	Matthäus, 80, 39, 267, no 404
872	Kly	Tomb 8	3	2	6	2	6	21	2	5	0	0	1	0	0	2.2	12	0	0	0	0	0	HerM	Matthäus, 80, 39, 145, no 177
869	Kly	0	3	2	6	2	6	21	2	2	0	0	0	0	0	0	0	0	0	0	0	0	HerM	Matthäus, 80, 39, 112, no 96
781	Kno	UM	2	2	2	2	5	21	2	3	0	0	3	0	0	15	65	0	0	0	0	0	StraM	Matthäus, 80, 10, 130, no 149
782	Kno	UM	2	2	2	2	5	21	2	21	0	0	3	0	0	0	0	0	0	0	0	0	StraM	Matthäus, 80, 10, 304, no 467
906	Sel	Tomb 4/I	2	2	6	2	6	21	2	2	0	0	3	0	0	0	25	0	0	0	0	0	StraM	Matthäus, 80, 40, 104, no 68
899	Sel	Tomb 3	2	2	6	2	6	21	2	17	0	0	3	0	0	0	23	0	0	0	0	0	StraM	Matthäus, 80, 40, 265, no 397
911	Sel	Tomb 4/I	2	2	6	2	6	21	2	9	0	0	3	0	0	35	18	0	0	0	0	0	StraM	Matthäus, 80, 40, 196, no 296
918	Sel	Tomb 4/I	2	2	6	2	6	21	2	9	0	0	2	0	0	31	9	0	0	0	0	0	StraM	Matthäus, 80, 40, 189, no 282
916	Sel	Tomb 4/I	2	2	6	2	6	21	2	17	0	0	3	0	0	14	20	0	0	0	0	0	StraM	Matthäus, 80, 40, 267, no 405
898	Sel	Tomb 3	2	2	6	2	6	21	2	3	0	0	3	0	0	0	0	0	0	0	0	0	StraM	Matthäus, 80, 40, 125, no 119
897	Sel	Tomb 3	2	2	6	2	6	21	2	3	0	0	3	0	0	0	30	0	0	0	0	0	StraM	Matthäus, 80, 40, 104, no 64
901	Sel	Tomb 3	2	2	6	2	6	21	2	20	0	0	3	0	0	4.4	15	0	0	0	0	0	StraM	Matthäus, 80, 40, 301, no 454
902	Sel	Tomb 3	2	2	6	2	6	21	2	20	0	0	3	0	0	0	0	0	0	0	0	0	StraM	Matthäus, 80, 40, 301, no 455
903	Sel	Tomb 3	2	2	6	2	6	21	2	23	0	0	3	0	0	0	0	0	0	0	0	0	StraM	Matthäus, 80, 40, 310, no 481
905	Sel	Tomb 4/I	2	2	6	2	6	21	2	2	0	0	3	0	0	0	33	0	0	0	0	0	StraM	Matthäus, 80, 40, 104, no 66
917	Sel	Tomb 4/I	2	2	6	2	6	21	2	20	0	0	1	0	0	2.9	15	0	0	0	0	0	StraM	Matthäus, 80, 40, 302, no 463
913	Sel	Tomb 4/I	2	2	6	2	6	21	2	11	0	0	2	0	0	5.6	29	0	0	0	0	0	StraM	Matthäus, 80, 40, 207, no 312
910	Sel	Tomb 4/I	2	2	6	2	6	21	2	9	0	0	2	0	0	22	8.9	0	0	0	0	0	StraM	Matthäus, 80, 40, 196, no 295
912	Sel	Tomb 4/I	2	2	6	2	6	21	2	10	0	0	3	0	0	0	0	0	0	0	0	0	StraM	Matthäus, 80, 40, 201, no 304
920	Sel	Tomb 4/III	2	2	6	2	6	21	2	12	0	0	1	0	0	3.4	15	0	0	0	0	0	StraM	Matthäus, 80, 40, 235, no 352
915	Sel	Tomb 4/I	2	2	6	2	6	21	2	17	0	0	1	0	0	15	22	0	0	0	0	0	StraM	Matthäus, 80, 40, 264, no 390
909	Sel	Tomb 4/I	2	2	6	2	6	21	2	9	0	0	2	0	0	22	8	0	0	0	0	0	StraM	Matthäus, 80, 40, 196, no 294
922	Sel	Tomb 4	2	2	6	2	6	21	2	2	0	0	3	0	0	0	0	0	0	0	0	0	StraM	Matthäus, 80, 40, 104, no 67
914	Sel	Tomb 4/I	2	2	6	2	6	21	2	12	0	0	1	0	0	4	14	0	0	0	0	0	StraM	Matthäus, 80, 40, 225, no 345
900	Sel	Tomb 3	2	2	6	2	6	21	2	18	0	0	2	0	0	4.5	15	0	0	0	0	0	StraM	Matthäus, 80, 40, 265, no 431
907	Sel	Tomb 4/I	2	2	6	2	6	21	2	3	0	0	3	0	0	0	30	0	0	0	0	0	StraM	Matthäus, 80, 40, 123, no 115
904	Sel	Tomb 4/I	2	2	6	2	6	21	2	2	0	0	3	0	0	0	36	0	0	0	0	0	StraM	Matthäus, 80, 40, 104, no 65
921	Sel	Tomb 4/III	2	2	6	2	6	21	2	20	0	0	2	0	0	3.8	14	0	0	0	0	0	StraM	Matthäus, 80, 40, 301, no 456
919	Sel	Tomb 4/III	2	2	6	2	6	21	2	1	0	0	1	0	0	8.8	30	0	0	0	0	0	StraM	Matthäus, 80, 40, 96, no 33
908	Sel	Tomb 4/I	2	2	6	2	6	21	2	3	0	0	2	0	0	0	44	0	0	0	0	0	StraM	Matthäus, 80, 40, 130, no 151
880	ZaP	Tomb 14	2	2	6	2	6	21	2	4	0	0	1	0	0	12	22	0	0	0	0	0	HerM	Matthäus, 80, 42, 141, no 164
894	ZaP	Tomb 36	2	2	6	2	6	21	2	17	0	0	1	0	0	15	26	0	0	0	0	0	1087	Matthäus, 80, 42, 267, no 406
895	ZaP	Tomb 99	2	2	6	2	6	21	2	1	0	0	2	0	0	6.7	25	0	65	9.1	0	1.1	AshM	Matthäus, 80, 42, 99, no 38
893	ZaP	Tomb 36	2	2	6	2	6	21	2	8	0	0	1	0	0	50	0	0	0	0	0	0	HerM	Matthäus, 80, 42, 172, no 242
896	ZaP	Tomb 99	2	2	6	2	6	21	2	9	0	0	2	0	0	9.5	0	0	0	0	0	0	HerM	Matthäus, 80, 42, 197, no 298
883	ZaP	Tomb 14	2	2	6	2	6	21	2	8	0	0	2	0	0	39	10	0	0	0	0	0	1088	Matthäus, 80, 42, 172, no 241
887	ZaP	Tomb 14	2	2	6	2	6	21	2	17	0	0	2	0	0	0	25	0	0	0	0	0	HerM	Matthäus, 80, 42, 266, no 398
890	ZaP	Tomb 14	2	2	6	2	6	21	2	19	0	0	1	0	0	5.9	12	0	0	0	0	0	1094	Matthäus, 80, 42, 297, no 449

Appendix V.2   Bronze objects. Sorted by object type / period / site. Total 1857

No	Site	Co	G	L	S	Pe	Da	Ot	C	TM	TK	TP	Cn	Le	Wi	He	Rd	We	Cu	Sn	As	Pb	Mu	Ref
882	ZaP	Tomb 14	2	2	6	2	6	21	2	5	0	0	1	0	0	2.6	11	0	0	0	0	0	1091	Matthäus, 80, 42, 145, no 178
884	ZaP	Tomb 14	2	2	6	2	6	21	2	10	0	0	2	0	0	34	7	0	0	0	0	0	1090	Matthäus, 80, 42, 201, no 303
891	ZaP	Tomb 14	2	2	6	2	6	21	2	20	0	0	1	0	0	3.8	16	0	0	0	0	0	1093	Matthäus, 80, 42, 302, no 464
881	ZaP	Tomb 14	2	2	6	2	6	21	2	4	0	0	1	0	0	0	20	0	0	0	0	0	HerM	Matthäus, 80, 42, 141, no 165
892	ZaP	Tomb 36	2	2	6	2	6	21	2	5	0	0	1	0	0	3.6	15	0	0	0	0	0	1092	Matthäus, 80, 42, 145, no 179
889	ZaP	Tomb 14	2	2	6	2	6	21	2	18	0	0	2	0	0	7.3	15	0	0	0	0	0	1528	Matthäus, 80, 42, 287, no 438
879	ZaP	Tomb 14	2	2	6	2	6	21	2	2	0	0	2	0	0	40	39	0	0	0	0	0	1081	Matthäus, 80, 42, 108, no 72
886	ZaP	Tomb 14	2	2	6	2	6	21	2	16	0	0	2	0	0	18	18	0	0	0	0	0	HerM	Matthäus, 80, 42, 259, no 368
885	ZaP	Tomb 14	2	2	6	2	6	21	2	11	0	0	2	0	0	5.8	34	0	0	0	0	0	1084	Matthäus, 80, 42, 212, no 329
878	ZaP	Tomb 14	2	2	6	2	6	21	2	1	1	0	2	0	0	12	42	0	0	0	0	0	1086	Matthäus, 80, 42, 97, no 37
888	ZaP	Tomb 14	2	2	6	2	6	21	2	18	0	0	2	0	0	5.5	27	0	0	0	0	0	1085	Matthäus, 80, 42, 285, no 433
854	Arh	Myc.B.E/6	2	2	6	3	7	21	2	0	0	0	3	0	0	0	0	0	0	0	0	0	HerM	Kallitsaki, 97, 215, 220
847	Arh	Myc.B.E/4	2	2	6	3	7	21	2	1	0	0	2	0	0	0	0	0	0	0	0	0	HerM	Matthäus, 80, 96, no29; Kallitsaki, 97, 220
852	Arh	Myc.B.E72	2	2	6	3	7	21	2	0	0	0	3	0	0	0	0	0	0	0	0	0	HerM	Kallitsaki, 97, 215, 220
858	Arh	Myc.B.E/6	2	2	6	3	7	21	2	0	0	0	3	0	0	0	0	0	0	0	0	0	HerM	Kallitsaki, 97, 215, 220
856	Arh	Myc.B.E/6	2	2	6	3	7	21	2	0	0	0	3	0	0	0	0	0	0	0	0	0	HerM	Kallitsaki, 97, 215, 220
853	Arh	Myc.B.E/5	2	2	6	3	7	21	2	0	0	0	2	0	0	0	0	0	0	0	0	0	HerM	Kallitsaki, 97, 215, 220
850	Arh	Myc.B.E/4	2	2	6	3	7	21	2	12	0	0	2	0	0	0	0	0	0	0	0	0	HerM	Sakellarakis / Sakellarakis, 97, 590; Kallitsaki, 97, 220
848	Arh	Myc.B.E/4	2	2	6	3	7	21	2	5	0	0	1	0	0	0	0	0	0	0	0	0	HerM	Matthäus, 80, 145, no 176; Kallitsaki, 97, 220
857	Arh	Myc.B.E/6	2	2	6	3	7	21	2	0	0	0	3	0	0	0	0	0	0	0	0	0	HerM	Kallitsaki, 97, 215, 220
859	Arh	Tholos B	2	2	6	3	7	21	2	20	0	0	1	0	0	0	0	0	0	0	0	0	HerM	Matthäus, 80, 302, no 462; Kallitsaki, 97, 220
855	Arh	Myc.B.E/6	2	2	6	3	7	21	2	0	0	0	3	0	0	0	0	0	0	0	0	0	HerM	Kallitsaki, 97, 215, 220
851	Arh	Myc.B.E/2	2	2	6	3	7	21	2	0	0	0	3	0	0	0	0	0	0	0	0	0	HerM	Kallitsaki, 97, 215, 220
849	Arh	Myc.B.E/6	2	2	6	3	7	21	2	17	0	0	1	0	0	0	0	0	0	0	0	0	HerM	Sakellarakis / Sakellarakis, 97, 590; Kallitsaki, 97, 220
867	Cha	0	4	1	6	3	0	21	2	16	0	0	3	0	0	0	0	0	0	0	0	0	ChaM	Matthäus, 80, 259, no 367
938	Cre	0	1	0	0	3	0	21	2	19	0	0	2	0	0	4.5	11	0	0	0	0	0	HNM	Matthäus, 80, 297, no 448, 298
931	EkP	Tomb A	2	2	6	3	0	21	2	17	0	0	1	0	0	15	0	0	0	0	0	0	2859	Matthäus, 80, 265, no 396
925	Gor	Tomb 2	2	2	6	3	8	21	2	14	0	0	1	0	0	11	13	0	0	0	0	0	2019	Matthäus, 80, 256, no 363
926	Gor	Tomb 2	2	2	6	3	8	21	2	19	0	0	2	0	0	1.8	5.9	0	0	0	0	0	2022	Matthäus, 80, 297, no 450, 298
1783	HaT	0	3	2	4	3	0	21	2	8	0	0	2	0	0	55	0	0	0	0	0	0	PigM	Evely, 93, 172, no 239
929	Iso	Tomb 4	2	2	6	3	7	21	2	11	0	0	2	0	0	5.2	27	0	0	0	0	0	1757	Matthäus, 80, 207, no 308
1476	Kla	0	4	1	6	3	8	21	2	0	0	0	1	0	0	0	0	0	0	0	0	0	0	Kanta, 80, 238
1839	Moc	Tomb 10	1	1	6	3	8	21	2	0	0	0	0	0	0	0	0	0	0	0	0	0	0	Soles / Davaras, 96, 212
1762	Moc	Tomb 10	1	1	6	3	0	21	2	0	0	0	0	0	0	0	13	0	0	0	0	0	0	Soles / Davaras, 96, 212, no CA 87
1838	Mol	Tomb A	1	1	6	3	9	21	2	9	0	0	2	0	31	8.5	0	0	0	0	0	0	1004	Matthäus, 80, 199, no 300
1390	Mon	0	4	2	6	3	7	21	2	0	0	0	1	0	0	0	0	0	0	0	0	0	0	Kanta, 80, 15
1452	Phk	0	1	2	6	3	7	21	2	0	0	0	1	0	0	0	0	0	0	0	0	0	0	Kanta, 80, 164, 183
933	Stm	Tomb1	2	2	6	3	0	21	2	18	0	0	1	0	0	0	15	0	0	0	0	0	HerM	Matthäus, 80, 285, no 432
1427	Tef	0	2	2	6	3	8	21	2	0	0	0	1	0	0	0	0	0	0	0	0	0	0	Kanta, 80, 80
1699	Gou	0	1	1	3	1	4	22	1	0	0	3	3	0	0	0	0	0	0	0	0	0	0	Boyd Hawes, 08, 34, no 63An

Appendix V.2 Bronze objects. Sorted by object type / period / site. Total 1857

No	Site	Co	G	L	S	Pe	Da	Ot	C	TM	TK	TP	Cn	Le	Wi	He	Rd	We	Cu	Sn	As	Pb	Mu	Ref
1694	Gou	N o Palace	1	1	3	1	4	22	1	0	0	0	3	0	0	0	14	0	0	0	0	0	608	Boyd Hawes, 08, 34, no 63A
1698	Gou	0	1	1	3	1	4	22	1	0	0	0	3	0	0	0	0	0	0	0	0	0	0	Boyd Hawes, 08, 34, no 63An
1701	Gou	Fg	1	1	3	1	4	22	1	0	0	0	3	2.1	0	0	0	0	0	0	0	0	611	Boyd Hawes, 08, 34, no 63B
1697	Gou	0	1	1	3	1	4	22	1	0	0	0	3	0	0	0	0	0	0	0	0	0	0	Boyd Hawes, 08, 34, no 63An
1696	Gou	0	1	1	3	1	4	22	1	0	0	0	3	0	0	0	0	0	0	0	0	0	0	Boyd Hawes, 08, 34, no 63An
1700	Gou	0	1	1	3	1	4	22	1	0	0	0	3	0	0	0	0	0	0	0	0	0	0	Boyd Hawes, 08, 34, no 63An
1695	Gou	0	1	1	3	1	4	22	1	0	0	0	3	0	0	0	0	0	0	0	0	0	0	Boyd Hawes, 08, 34, no 63An
1847	Moc	House C3	1	1	3	1	4	22	1	0	0	0	3	0	0	0	0	0	0	0	0	0	0	Soles / Davaras, 96, 194, no CA 49
1743	Moc	House C3	1	1	3	1	4	22	1	0	0	0	3	0	0	0	0	0	0	0	0	0	0	Soles / Davaras, 96, 194, no CA 50
1591	Kom	N House	3	1	3	0	0	29	1	0	0	0	3	2.5	2.4	0	0	0	0	0	0	0	0	Blitzer, 95, 515, no M 142
1590	Kom	0	3	1	3	0	0	29	1	0	0	0	1	3.2	1.7	0	0	0	0	0	0	0	0	Blitzer, 95, 515, no M 127
1589	Kom	0	3	1	3	0	0	29	1	0	0	0	2	13	0.1	0	0	0	0	0	0	0	0	Blitzer, 95, 514, no M 106
1586	Kom	0	3	1	3	0	0	29	1	0	0	0	3	1.3	0.3	0	0	0	0	0	0	0	0	Blitzer, 95, 512, no M 71
1583	Kom	0	3	1	3	0	0	29	1	0	0	0	1	3.3	1.7	0	0	0	0	0	0	0	0	Blitzer, 95, 511, no M 60
1665	Gou	0	1	1	3	1	4	29	1	0	0	0	1	0	0	0	0	0	0	0	0	0	623	Boyd Hawes, 08, 34, no 64
1662	Gou	0	1	1	3	1	4	29	1	0	0	0	1	7.2	0	0	0	0	0	0	0	0	621	Boyd Hawes, 08, 34, no 45
1661	Gou	0	1	1	3	1	4	29	1	0	0	0	1	0	0	0	0	0	0	0	0	0	0	Boyd Hawes, 08, 34, no 44n
1663	Gou	0	1	1	3	1	4	29	1	0	0	0	1	2.6	0	0	0	0	0	0	0	0	0	Boyd Hawes, 08, 34, no 46
1664	Gou	0	1	1	3	1	4	29	1	0	0	0	1	0	0	0	0	0	0	0	0	0	0	Boyd Hawes, 08, 34, no 46n
1660	Gou	0	1	1	3	1	4	29	1	0	0	0	2	0	0	0	0	0	0	0	0	0	617	Boyd Hawes, 08, 34, no 44
1588	Kom	0	3	1	3	1	0	29	1	0	0	0	3	1.7	1	0	0	0	0	0	0	0	0	Blitzer, 95, 513, no M 93
1730	Moc	House A2	1	1	3	1	4	29	1	0	0	0	1	0	0	0	0	0	0	0	0	0	0	Soles / Davaras, 94, 419, no CA 24
1848	Zak	Obliq. Build1	1	1	2	1	8	29	1	0	0	0	1	4	0.2	0	0	0	0	0	0	0	662	Platon, 88, II, 223, no 8
125	Zak	Obliq. Build1	1	1	2	1	7	29	1	0	0	0	1	5	0.2	0	0	0	0	0	0	0	662	Platon, 88, II, 223, no 7
119	Zak	Harb. Road	1	1	2	1	0	29	1	0	0	0	1	5	0	0	0	0	0	0	0	0	822	Platon, 88, II, 222, no 1
124	Zak	Obliq. Build1	1	1	2	1	6	29	1	0	0	0	1	6	0.3	0	0	0	0	0	0	0	662	Platon, 88, II, 223, no 6
123	Zak	Buil. Tu	1	1	2	1	0	29	1	0	0	0	1	3.5	0.3	0	0	0	0	0	0	0	0	Platon, 88, II, 223, no 5
121	Zak	XXVIII	1	1	1	1	4	29	1	0	0	0	1	3.5	0.2	0	0	0	0	0	0	0	0	Platon, 88, II, 223, no 3
122	Zak	Buil. SD	1	1	2	1	0	29	1	0	0	0	1	3	0.2	0	0	0	0	0	0	0	0	Platon, 88, II, 223, no 4
126	Zak	Obliq. Build1	1	1	2	1	0	29	1	0	0	0	1	3.5	0.2	0	0	0	0	0	0	0	662	Platon, 88, II, 223, no 9
120	Zak	Strong Bui.	1	1	2	1	0	29	2	0	0	0	2	4	0	0	0	0	0	0	0	0	820	Platon, 88, II, 223, no 2
1585	Kom	0	3	1	3	2	5	29	1	0	0	0	3	2.3	0.3	0	0	0	0	0	0	0	0	Blitzer, 95, 512, no M 70
1584	Kom	0	3	1	3	2	5	29	1	0	0	0	3	1.8	0	0	0	0	0	0	0	0	0	Blitzer, 95, 512, no M 69
1587	Kom	0	3	1	3	2	5	29	1	0	0	0	3	2.7	0.3	0	0	0	0	0	0	0	0	Blitzer, 95, 512, no M 72
1188	Sel	Tomb 4/I	2	2	6	2	6	29	1	0	0	0	1	7	0	0	0	0	0	20	0	0	0	Catling/Catling, 74, 230, no12;Catling/Jones, 76, 22,no8
1187	Sel	Tomb 4/I	2	2	6	2	6	29	1	0	0	0	1	0	0	10	0	0	0	8.5	0	0	0	Catling/Catling, 74, 229, no11;Catling/Jones, 76, 22,no6
1592	Kom	Cliffside H	3	1	3	3	0	29	1	0	0	0	2	2	9	0	0	0	0	0	0	0	0	Blitzer, 95, 516, no M 145
758	Cre	0	1	0	0	0	0	31	2	0	14	0	3	21	0	0	0	0	0	0	0	0	HNM	Kilian-Dirlmeier 93, 83, no 184
581	Ark	0	2	2	7	1	0	31	3	0	1	0	2	53	0	0	0	0	0	0	0	0	HerM	Kilian-Dirlmeier, 93, 11, no 12
582	Ark	0	2	2	7	1	0	31	3	0	1	0	3	42	0	0	0	0	0	0	0	0	HerM	Kilian-Dirlmeier, 93, 12, no 13

Appendix V.2 Bronze objects. Sorted by object type / period / site. Total 1857

No	Site	Co	G	L	S	Pe	Da	Ot	C	TM	TK	TP	Cn	Le	Wi	He	Rd	We	Cu	Sn	As	Pb	Mu	Ref
584	Ark	0	2	2	7	1	0	31	3	0	1	0	2	35	0	0	0	0	0	0	0	0	HerM	Kilian-Dirlmeier, 93, 12, no 14A
595	Ark	0	2	2	7	1	0	31	3	0	2	0	1	100	0	0	0	0	0	0	0	0	HerM	Kilian-Dirlmeier, 93, 18, no 30, 14
590	Ark	0	2	2	7	1	0	31	3	0	2	0	1	102	0	0	0	0	0	0	0	0	2402	Kilian-Dirlmeier, 93, 18, no 25, 14
580	Ark	0	2	2	7	1	0	31	3	0	1	0	2	102	0	0	0	0	0	0	0	0	2403	Kilian-Dirlmeier, 93, 11, no 9
583	Ark	0	2	2	7	1	0	31	3	0	1	0	3	35	0	0	0	0	0	0	0	0	HerM	Kilian-Dirlmeier, 93, 12, no 14
588	Ark	0	2	2	7	1	0	31	3	0	2	0	1	106	0	0	0	0	0	0	0	0	2400	Kilian-Dirlmeier, 93, 18, no 23, 14
587	Ark	0	2	2	7	1	0	31	3	0	2	0	1	111	0	0	0	0	0	0	0	0	2399	Kilian-Dirlmeier, 93, 18, no 22, 14
589	Ark	0	2	2	7	1	0	31	3	0	2	0	1	104	0	0	0	0	0	0	0	0	2401	Kilian-Dirlmeier, 93, 18, no 24, 14
592	Ark	0	2	2	7	1	0	31	3	0	2	0	1	100	0	0	0	0	0	0	0	0	2817	Kilian-Dirlmeier, 93, 18, no 27, 14
594	Ark	0	2	2	7	1	0	31	3	0	2	0	1	100	0	0	0	0	0	0	0	0	HerM	Kilian-Dirlmeier, 93, 18, no 29, 14
585	Ark	0	2	2	7	1	0	31	3	0	1	0	3	31	0	0	0	0	0	0	0	0	HerM	Kilian-Dirlmeier, 93, 12, no 14B
591	Ark	0	2	2	7	1	0	31	3	0	2	0	1	100	0	0	0	0	0	0	0	0	2403	Kilian-Dirlmeier, 93, 18, no 26, 14
593	Ark	0	2	2	7	1	0	31	3	0	2	0	1	100	0	0	0	0	0	0	0	0	HerM	Kilian-Dirlmeier, 93, 18, no 28, 14
586	Ark	0	2	2	7	1	0	31	3	0	1	0	1	82	0	0	0	1800	0	0	0	0	2404	Kilian-Dirlmeier, 93, 12, no 15
613	Iso	Iso Dep	2	2	6	1	0	31	2	0	15	0	3	0	0	0	0	0	0	0	0	0	0	Kilian-Dirlmeier, 93, 18, no 216
612	Iso	Iso Dep	2	2	6	1	0	31	2	0	2	0	3	6	0	0	0	0	0	0	0	0	0	Kilian-Dirlmeier, 93, 18, no 21
599	Mal	Quart. III	2	1	1	2	0	31	2	0	2	0	2	74	0	0	0	0	0	0	0	0	2285	Kilian-Dirlmeier, 93, 18, no 33, 26
598	Mal	Quart. III	2	1	1	2	0	31	2	0	2	0	2	81	0	0	0	0	0	0	0	0	2284	Kilian-Dirlmeier, 93, 18, no 32, 26
1354	Pak	Block Chi	1	1	3	4	0	31	2	0	0	0	1	40	0	0	0	0	0	0	0	0	0	Dawkins, 04-05, 282
641	PaZ	0	1	2	0	1	0	31	2	0	2	0	2	85	0	0	0	0	0	0	0	0	2505	Kilian-Dirlmeier, 93, 18, no 34
597	Psy	0	1	2	7	1	0	31	3	0	1	0	2	57	0	0	0	0	0	0	0	0	AshM	Kilian-Dirlmeier, 93, 11, no 11, 14
596	Psy	0	1	2	7	1	0	31	3	0	1	0	2	55	0	0	0	0	0	0	0	0	AshM	Kilian-Dirlmeier, 93, 11, no10, 14
640	Ska	0	1	2	5	1	0	31	2	0	2	0	3	21	0	0	0	0	0	0	0	0	AshM	Kilian-Dirlmeier, 93, 18, no 20, 26
601	Zak	XII	1	1	1	1	4	31	2	0	2	0	2	92	0	0	0	0	0	0	0	0	2590	Kilian-Dirlmeier, 93, 18, no 36
600	Zak	XII	1	1	1	1	4	31	2	0	2	0	2	84	0	0	0	0	0	0	0	0	2591	Kilian-Dirlmeier, 93, 18, no 35
614	AgC	0	2	2	6	2	0	31	2	0	5	0	1	61	0	0	0	0	0	0	0	0	HerM	Kilian-Dirlmeier, 93, 42, no 56
615	Akr	0	2	2	6	2	0	31	2	0	5	0	2	65	0	0	0	0	0	0	0	0	AshM	Kilian-Dirlmeier, 93, 44, no 60, 52
1711	Arh	Bur.Buil. 3	2	2	6	2	0	31	2	0	6	0	1	46	0	0	0	0	0	0	0	0	HerM	Sakellarakis / Sakellarakis, 97, 599
644	Arh	Tomb B	2	2	6	2	0	31	2	0	5	0	1	56	0	0	0	0	0	0	0	0	2899	Kilian-Dirlmeier, 93, 46, no 77
630	Arm	Tomb 200	4	2	6	2	0	31	2	0	11	0	1	72	0	0	0	0	0	0	0	0	0	Papadopoulou, 97, 331-333,
629	Arm	Tomb 35	4	2	6	2	0	31	2	0	11	0	1	37	0	0	0	0	0	0	0	0	ChaM	Kilian-Dirlmeier, 93, 60, no 125
631	Cha	Region 1	4	1	6	2	0	31	2	0	6	0	2	38	0	0	0	0	0	0	0	0	ChaM	Kilian-Dirlmeier, 93, 46, no 74; Andreadaki, 97, 488
632	Cha	Region 1	4	1	6	2	0	31	2	0	11	0	2	43	0	0	0	0	0	0	0	0	ChaM	Kilian-Dirlmeier, 93, 62, no 137;Andreadaki, 97, 488
643	Che	0	2	1	0	2	0	31	2	0	5	0	2	33	0	0	0	0	0	0	0	0	GiaC	Kilian-Dirlmeier, 93, 45, no 65
618	Hal	Tomb 2	2	2	6	2	5	31	2	0	11	0	1	62	0	0	0	0	0	0	0	0	2497	Kilian-Dirlmeier, 93, 58, no 111
619	Hal	Tomb 1	2	2	6	2	5	31	2	0	11	0	2	40	0	0	0	0	0	0	0	0	2372	Kilian-Dirlmeier, 93, 59, no 112
651	Her	0	2	1	0	2	6	31	2	0	13	0	2	25	0	0	0	0	0	0	0	0	2580	Kilian-Dirlmeier, 93, 82, no 173, 87
642	Kly	Tomb 8	3	2	6	2	6	31	2	0	5	0	2	42	0	0	0	0	0	0	0	0	690	Kilian-Dirlmeier, 93, 44, no 57; Kanta 80, 99
625	Kno	NW House	2	2	1	2	2	31	2	0	12	0	2	37	0	0	0	0	0	0	0	0	HerM	Kilian-Dirlmeier, 93, 64, no 154; Georgiou, 79, 20
624	Mas	Tomb XVIII	2	2	6	2	6	31	2	0	11	0	1	46	0	0	0	0	0	0	0	0	2141	Kilian-Dirlmeier, 93, 62, no 142, 66

Appendix V.2 Bronze objects. Sorted by object type / period / site. Total 1857

No	Site	Co	G	L	S	Pe	Da	Ot	C	TM	TK	TP	Cn	Le	Wi	He	Rd	We	Cu	Sn	As	Pb	Mu	Ref
621	NeH	Tomb V	2	2	6	2	6	31	2	0	11	0	1	47	0	0	0	0	0	0	0	0	2375	Kilian-Dirlmeier, 93, 59, no 115
620	NeH	Tomb II	2	2	6	2	6	31	2	0	11	0	1	58	0	0	0	0	0	0	0	0	710	Kilian-Dirlmeier, 93, 59, no 113
652	Nek	0	4	2	4	2	6	31	2	0	14	0	1	45	0	0	0	0	0	0	0	0	ChaM	Kilian-Dirlmeier, 93, 82, no 174
647	Pig	0	4	2	6	2	6	31	2	0	11	0	2	43	0	0	0	0	0	0	0	0	ChaM	Kilian-Dirlmeier, 93, 59, no 116; Kanta 80, 213
623	Sel	Tomb 4/II	2	2	6	2	6	31	2	0	11	0	2	70	0	0	0	0	0	10	1	0	HerM	Kilian-Dirlmeier, 93, 60, no 122; Catling et al, 76, 22
622	Sel	Tomb 4/I	2	2	6	2	6	31	2	0	11	0	1	47	0	0	0	0	0	0	0	0	2973	Kilian-Dirlmeier, 93, 60, no 121
608	ZaP	Tomb 55	2	2	6	2	6	31	2	0	11	0	1	66	0	0	0	0	0	0	0	0	AshM	Kilian-Dirlmeier, 93, 60, no120; Baboula / N., 99, 146
603	ZaP	Tomb 36	2	2	6	2	6	31	2	0	5	0	1	96	0	0	0	0	0	0	0	0	1097	Kilian-Dirlmeier, 93, 43, no 52
604	ZaP	Tomb 44	2	2	6	2	6	31	2	0	5	0	2	82	0	0	0	0	0	0	0	0	AshM	Kilian-Dirlmeier, 93, 44, no 53; Baboula / N., 99, 146
602	ZaP	Tomb 44	2	2	6	2	6	31	2	0	3	0	2	61	0	0	0	0	0	0	0	0	1456a	Kilian-Dirlmeier, 93, 35, no 45
610	ZaP	Tomb 36	2	2	6	2	6	31	2	0	11	0	1	61	0	0	0	0	0	0	0	0	1098	Kilian-Dirlmeier, 93, 60, no 124
657	Adr	0	1	2	6	3	0	31	2	0	14	0	2	26	0	0	0	0	0	0	0	0	HerM	Kilian-Dirlmeier, 93, 84, no 196
616	Akr	0	2	2	6	3	0	31	2	0	12	0	3		0	0	0	0	0	0	0	0	AshM	Kilian-Dirlmeier, 93, 64, no 152
633	Cha	0	4	1	0	3	0	31	2	0	0	0	3	0	0	0	0	0	0	0	0	0	ChaM	Kilian-Dirlmeier, 93, 93, no 219
663	Cre	0	0	0	0	3	0	31	2	0	11	0	3	21	0	0	0	0	0	0	0	0	GiaC	Kilian-Dirlmeier, 93, 59, no 117
665	Cre	0	0	0	0	3	0	31	2	0	13	0	2	32	0	0	0	0	0	0	0	0	GiaC	Kilian-Dirlmeier, 93, 82, no 172
661	Cre	0	0	0	0	3	0	31	2	0	7	0	2	20	0	0	0	0	0	0	0	0	Kass	Kilian-Dirlmeier, 93, 48, no 87
662	Cre	0	0	0	0	3	0	31	2	0	11	0	3	21	0	0	0	0	0	0	0	0	409	Kilian-Dirlmeier, 93, 58, no 110
666	Cre	0	1	0	0	3	0	31	2	0	7	0	2	40	0	0	0	0	0	0	0	0	HNM	Kilian-Dirlmeier, 93, 48, no 91
664	Cre	0	0	0	0	3	0	31	2	0	11	0	2	37	0	0	0	0	0	0	0	0	GiaC	Kilian-Dirlmeier, 93, 62, no 141
653	EkP	0	2	2	6	3	0	31	2	0	14	0	1	35	0	0	0	0	0	0	0	0	HerM	Kilian-Dirlmeier, 93, 82, no 179
650	Gor	Tomb 2	2	2	6	3	8	31	2	0	12	0	2	34	0	0	0	0	0	0	0	0	2025	Kilian-Dirlmeier, 93, 63, no 150; Kanta, 80, 48
660	Gra	0	0	0	0	3	0	31	2	0	16	0	1	87	0	0	0	0	0	0	0	0	AshM	Kilian-Dirlmeier, 93, 97, no 246
659	Gra	0	0	0	0	3	0	31	2	0	16	0	1	85	0	0	0	0	0	0	0	0	AshM	Kilian-Dirlmeier, 93, 97, no 245
655	HaG	0	1	2	6	3	8	31	2	0	14	0	1	33	0	0	0	0	0	0	0	0	2429	Kilian-Dirlmeier, 93, 83, no 190
617	Hal	Tomb 3	2	2	6	3	8	31	2	0	7	0	1	83	0	0	0	0	0	0	0	0	494	Kilian-Dirlmeier, 93, 48, no 88
648	HaS	Tomb I/B	2	2	6	3	0	31	2	0	11	0	1	51	0	0	0	0	0	0	0	0	4467	Kilian-Dirlmeier, 93, 62, no 136
654	Kma	0	1	2	6	3	8	31	2	0	14	0	2	24	0	0	0	0	0	0	0	0	HNM	Kilian-Dirlmeier, 93, 83, no 189; Kanta, 80, 161
658	Mes	0	3	2	0	3	0	31	2	0	16	0	1	54	0	0	0	0	0	0	0	0	641	Kilian-Dirlmeier, 93, 97, no 241
1834	Mol	Tomb A	1	1	6	3	9	31	2	0	16	0	3	27	0	0	0	0	0	0	0	0	999	Kilian-Dirlmeier, 93, 97, no 247
1832	Mol	Tomb B	1	1	6	3	9	31	2	0	16	0	2	41	0	0	0	0	0	0	0	0	1011	Kilian-Dirlmeier, 93, 95, no 230
1831	Mol	Tomb A	1	1	6	3	9	31	2	0	14	0	2	44	0	0	0	0	0	0	0	0	998	Kilian-Dirlmeier, 93, 85, no 205
1830	Mol	Tomb A	1	1	6	3	9	31	2	0	14	0	1	58	0	0	0	0	0	0	0	0	997	Kilian-Dirlmeier, 93, 85, no 204
1833	Mol	Tomb B	1	1	6	3	9	31	2	0	16	0	2	51	0	0	0	0	0	0	0	0	1010	Kilian-Dirlmeier, 93, 97, no 242
634	Myr	Tomb B	1	1	6	3	0	31	2	0	11	0	1	31	0	0	0	0	0	0	0	0	2741	Kilian-Dirlmeier, 93, 62, no 140
636	Myr	Tomb A	1	1	6	3	0	31	2	0	16	0	2	51	0	0	0	0	0	0	0	0	2745	Kilian-Dirlmeier, 93, 95, no 227
635	Myr	Tomb B	1	1	6	3	0	31	2	0	14	0	2	45	0	0	0	0	0	0	0	0	2739	Kilian-Dirlmeier, 93, 82, no 175
649	Pak	Beehive T	1	1	6	3	7	31	2	0	12	0	3	10	0	0	0	0	0	0	0	0	1365	Kilian-Dirlmeier, 93, 63, no 149; Sandars, 63, 148
645	Phk	0	2	2	6	3	7	31	2	0	7	0	1	38	0	0	0	0	0	0	0	0	HNM	Kilian-Dirlmeier, 93, 48, no 90
637	Sit	0	1	1	0	3	0	31	2	0	14	0	1	51	0	0	0	0	0	0	0	0	AshM	Kilian-Dirlmeier, 93, 84, no 203

Appendix V.2 Bronze objects. Sorted by object type / period / site. Total 1857

No	Site	Co	G	L	S	Pe	Da	Ot	C	TM	TK	TP	Cn	Le	Wi	He	Rd	We	Cu	Sn	As	Pb	Mu	Ref
638	Sit	0	1	1	0	3	0	31	2	0	16	0	2	51	0	0	0	0	0	0	0	0	AshM	Kilian-Dirlmeier, 93, 97, no 240
639	Sit	0	1	1	0	3	0	31	2	0	16	0	2	56	0	0	0	0	0	0	0	0	ChaM	Kilian-Dirlmeier, 93, 99, no 260
646	Sit	0	1	1	0	3	0	31	2	0	7	0	1	50	0	0	0	0	0	0	0	0	FwM	Kilian-Dirlmeier, 93, 48, no 96
656	Stc	0	1	2	0	3	8	31	2	0	14	0	3	0	0	0	0	0	0	0	0	0	2231	Kilian-Dirlmeier, 93, 83, no 193
757	Stm	Tomb E	2	2	6	3	7	31	2	0	14	0	1	37	0	0	0	0	0	0	0	0	0	Kilian-Dirlmeier, 93, 82, no 177
627	Sym	E	1	2	8	3	0	31	2	0	7	0	1	103	0	0	0	0	0	0	0	0	HerM	Kilian-Dirlmeier, 93, 48, no 89
628	Sym	E	1	2	8	3	0	31	2	0	7	0	1	61	0	0	0	0	0	0	0	0	HerM	Kilian-Dirlmeier, 93, 48, no 94
626	Sym	E	1	2	8	3	0	31	2	0	7	0	1	104	0	0	0	0	0	0	0	0	3822	Kilian-Dirlmeier, 93, 47, no 85
606	ZaP	Tomb 98	2	2	6	3	0	31	2	0	11	0	1	61	0	0	0	0	82.7	12.2	0.4	0.1	1100	Kilian-Dirlmeier, 93, 59, no 114; Mangou / I, 98, 94
609	ZaP	Tomb 42	2	2	6	3	0	31	2	0	11	0	1	62	0	0	0	0	82.9	11.1	0.3	0	1099	Kilian-Dirlmeier, 93, 60, no 123; Mangou / I, 98, 94
605	ZaP	Tomb 14	2	2	6	3	0	31	2	0	7	0	1	61	0	0	0	0	0	0	0	0	1102	Kilian-Dirlmeier, 93, 48, no 93
611	ZaP	Tomb 95	2	2	6	3	0	31	2	0	14	0	1	37	0	0	0	0	0	0	0	0	1103	Kilian-Dirlmeier, 93, 84, no 195
607	ZaP	Tomb 43	2	2	6	3	0	31	2	0	11	0	1	50	0	0	0	0	0	0	0	0	1101	Kilian-Dirlmeier, 93, 60, no 119
1361	Cre	0	0	0	0	0	0	32	2	0	0	2	2	32	0	0	0	0	0	0	0	0	GiaC	Driessen / Macdonald, 84, 73, no Ei 4
1360	Cre	0	0	0	0	0	0	32	2	0	0	2	1	32	0	0	0	0	0	0	0	0	GiaC	Driessen / Macdonald, 84, 73, no Ei 3
1569	Pat	0	4	2	7	0	0	32	2	0	0	1	1	19	0	0	0	0	94.9	1.8	1.6	0.1	AshM	Boardman, 61, 78, no 376, 160
1343	Psy	0	1	2	7	0	0	32	2	0	0	0	3	9.6	0	0	0	0	0	0	0	0	468	Boardman, 61,17
1645	Gou	0	1	1	3	1	1	32	2	0	0	1	1	27	0	0	0	0	86.9	13.1	0	0	CaM	Charles, 68, 278
1639	Gou	0	1	1	3	1	4	32	2	0	0	0	1	0	0	0	0	0	0	0	0	0	0	Boyd Hawes, 08, 34, no 58n
1615	Gou	Hill House	1	1	3	1	4	32	2	0	0	1	2	37	0	0	0	0	0	0	0	0	930	Boyd Hawes, 08, 34, no 49
1628	Gou	0	1	1	3	1	4	32	2	0	0	0	1	0	0	0	0	0	0	0	0	0	0	Boyd Hawes, 08, 34, no 57n
1620	Gou	0	1	1	3	1	4	32	2	0	0	0	1	0	0	0	0	0	0	0	0	0	0	Boyd Hawes, 08, 34, no 53n
1623	Gou	0	1	1	3	1	4	32	2	0	0	0	1	0	0	0	0	0	0	0	0	0	0	Boyd Hawes, 08, 34, no 54n
1636	Gou	0	1	1	3	1	4	32	2	0	0	0	1	0	0	0	0	0	0	0	0	0	0	Boyd Hawes, 08, 34, no 58n
1846	Gou	Hf	1	1	3	1	4	32	2	0	0	0	1	14	0	0	0	0	0	0	0	0	0	Boyd Hawes, 08, 34, no 61
1637	Gou	0	1	1	3	1	4	32	2	0	0	0	1	0	0	0	0	0	0	0	0	0	0	Boyd Hawes, 08, 34, no 58n
1640	Gou	0	1	1	3	1	4	32	2	0	0	0	1	22	0	0	0	0	0	0	0	0	576	Boyd Hawes, 08, 34, no 59
1644	Gou	0	1	1	3	1	4	32	2	0	0	0	1	22	0	0	0	0	0	0	0	0	575	Boyd Hawes, 08, 34, no 60
1616	Gou	F 14	1	1	3	1	4	32	2	0	0	0	1	36	0	0	0	0	0	0	0	0	574	Boyd Hawes, 08, 34, no 50
1630	Gou	0	1	1	3	1	4	32	2	0	0	0	1	0	0	0	0	0	0	0	0	0	0	Boyd Hawes, 08, 34, no 58n
1642	Gou	0	1	1	3	1	4	32	2	0	0	0	1	0	0	0	0	0	0	0	0	0	0	Boyd Hawes, 08, 34, no 59n
1635	Gou	0	1	1	3	1	4	32	2	0	0	0	1	0	0	0	0	0	0	0	0	0	0	Boyd Hawes, 08, 34, no 58n
1633	Gou	0	1	1	3	1	4	32	2	0	0	0	1	0	0	0	0	0	0	0	0	0	0	Boyd Hawes, 08, 34, no 58n
1631	Gou	0	1	1	3	1	4	32	2	0	0	0	1	0	0	0	0	0	0	0	0	0	0	Boyd Hawes, 08, 34, no 58n
1622	Gou	0	1	1	3	1	4	32	2	0	0	0	1	0	0	0	0	0	0	0	0	0	0	Boyd Hawes, 08, 34, no 54n
1643	Gou	0	1	1	3	1	4	32	2	0	0	0	1	0	0	0	0	0	0	0	0	0	0	Boyd Hawes, 08, 34, no 59n
1632	Gou	0	1	1	3	1	4	32	2	0	0	0	1	0	0	0	0	0	0	0	0	0	0	Boyd Hawes, 08, 34, no 58n
1626	Gou	Hill House	1	1	3	1	4	32	2	0	0	0	1	21	0	0	0	0	0	0	0	0	577	Boyd Hawes, 08, 34, no 56
1618	Gou	0	1	1	3	1	4	32	2	0	0	0	1	18	0	0	0	0	0	0	0	0	589	Boyd Hawes, 08, 34, no 52
1625	Gou	0	1	1	3	1	4	32	2	0	0	0	1	0	0	0	0	0	0	0	0	0	0	Boyd Hawes, 08, 34, no 55n

Appendix V.2 Bronze objects. Sorted by object type / period / site. Total 1857

No	Site	Co	G	L	S	Pe	Da	Ot	C	TM	TK	TP	Cn	Le	Wi	He	Rd	We	Cu	Sn	As	Pb	Mu	Ref
1634	Gou	0	1	1	3	1	4	32	2	0	0	1	1	0	0	0	0	0	0	0	0	0	0	Boyd Hawes, 08, 34, no 58n
1641	Gou	0	1	1	3	1	4	32	2	0	0	1	1	0	0	0	0	0	0	0	0	0	0	Boyd Hawes, 08, 34, no 59n
1638	Gou	0	1	1	3	1	4	32	2	0	0	1	1	0	0	0	0	0	0	0	0	0	0	Boyd Hawes, 08, 34, no 58n
1621	Gou	0	1	1	3	1	4	32	2	0	0	1	1	12	0	0	0	0	0	0	0	0	590	Boyd Hawes, 08, 34, no 54
1629	Gou	Hill House	1	1	3	1	4	32	2	0	0	1	1	16	0	0	0	0	0	0	0	0	579	Boyd Hawes, 08, 34, no 58
1627	Gou	Hill House	1	1	3	1	4	32	2	0	0	1	1	16	0	0	0	0	0	0	0	0	578	Boyd Hawes, 08, 34, no 57
1619	Gou	0	1	1	3	1	4	32	2	0	0	1	1	17	0	0	0	0	0	0	0	0	583	Boyd Hawes, 08, 34, no 53
1617	Gou	0	1	1	3	1	4	32	2	0	0	1	1	17	0	0	0	0	0	0	0	0	588	Boyd Hawes, 08, 34, no 51
1624	Gou	0	1	1	3	1	4	32	2	0	0	1	1	13	0	0	0	0	0	0	0	0	586	Boyd Hawes, 08, 34, no 55
1793	HaT	Room 27	3	2	4	1	4	32	2	0	0	0	1	21	0	0	0	0	0	0	0	0	0	Halbherr et.al., 80, 41
1796	HaT	Casa Est	3	2	4	1	3	32	2	0	0	0	1	30	0	0	0	0	0	0	0	0	0	Halbherr et.al., 80, 287
1795	HaT	Room 69	3	2	4	1	4	32	2	0	0	0	2	0	0	0	0	0	0	0	0	0	0	Halbherr et.al., 80, 167
1828	HaT	Casa Est	3	2	4	1	3	32	2	0	0	0	2	20	0	0	0	0	0	0	0	0	0	Halbherr et.al., 80, 287
1794	HaT	0	3	2	4	1	4	32	2	0	0	0	2	21	0	0	0	0	0	0	0	0	728	Halbherr et.al., 80, 68
1490	Kno	NW House	2	2	2	1	0	32	2	0	0	0	1	27	0	0	0	0	0	0	0	0	0	Georgiou, 79, 20, no 11
1491	Kno	NW House	2	2	2	1	0	32	2	0	0	0	1	26	0	0	0	0	0	0	0	0	0	Georgiou, 79, 20, no 12
1489	Kno	NW House	2	2	2	1	0	32	2	0	0	0	1	25	0	0	0	0	0	0	0	0	0	Georgiou, 79, 20, no 10
1488	Kno	NW House	2	2	2	1	0	32	2	0	0	0	1	17	0	0	0	0	0	0	0	0	0	Georgiou, 79, 20, no 9
1499	Mal	XXV 2	2	2	1	1	0	32	2	0	0	1	1	14	0	0	0	0	0	0	0	0	2098	Georgiou, 79, 37, no 9
1500	Mal	XXV 2	2	2	1	1	0	32	2	0	0	1	1	11	0	0	0	0	0	0	0	0	2099	Georgiou, 79, 37, no 10
1498	Mal	XXV 2	2	2	1	1	0	32	2	0	0	1	1	15	0	0	0	0	0	0	0	0	2097	Georgiou, 79, 37, no 8
1815	Mal	Quart. E	2	2	2	1	3	32	2	0	0	0	1	0	0	0	0	0	0	0	0	0	0	Driessen / Macdonald, 97, 191
1813	Mal	Quart. E	2	2	2	1	3	32	2	0	0	0	1	0	0	0	0	0	0	0	0	0	0	Driessen / Macdonald, 97, 191
1816	Mal	Quart. E	2	2	2	1	3	32	2	0	0	0	1	0	0	0	0	0	0	0	0	0	0	Driessen / Macdonald, 97, 191
1814	Mal	Quart. E	2	2	2	1	3	32	2	0	0	0	1	0	0	0	0	0	0	0	0	0	0	Driessen / Macdonald, 97, 191
1807	Mal	Quat.Z, Zg	2	2	2	1	0	32	2	0	0	0	1	0	0	0	0	0	0	0	0	0	0	Driessen / Macdonald, 97, 190
1812	Mal	Quart. E	2	2	2	1	3	32	2	0	0	0	1	0	0	0	0	0	0	0	0	0	0	Driessen / Macdonald, 97, 191
1811	Mal	Quart. E	2	2	2	1	0	32	2	0	0	0	1	0	0	0	0	0	0	0	0	0	0	Driessen / Macdonald, 97, 191
1804	Mal	Quat.Z, Zg	2	2	2	1	0	32	2	0	0	0	1	0	0	0	0	0	0	0	0	0	0	Driessen / Macdonald, 97, 190
1808	Mal	Quart.Z, Zg	2	2	2	1	0	32	2	0	0	0	1	0	0	0	0	0	0	0	0	0	0	Driessen / Macdonald, 97, 190
1810	Mal	Quart.Z, Zg	2	2	2	1	0	32	2	0	0	0	1	0	0	0	0	0	0	0	0	0	0	Driessen / Macdonald, 97, 190
1809	Mal	Quart.Z, Zg	2	2	2	1	0	32	2	0	0	0	1	0	0	0	0	0	0	0	0	0	0	Driessen / Macdonald, 97, 190
1501	Mal	XXV 2	2	2	1	1	0	32	2	0	1	0	3	0	0	0	0	0	0	0	0	0	0	Georgiou, 79, 37, no 11
1502	Mal	XXV 2	2	2	1	1	0	32	2	0	1	0	3	0	0	0	0	0	0	0	0	0	0	Georgiou, 79, 37, no 11
1805	Mal	Quart.Z, Zg	2	2	2	1	0	32	2	0	0	0	1	0	0	0	0	0	0	0	0	0	0	Driessen / Macdonald, 97, 190
1806	Mal	Quat.Z, Zg	2	2	2	1	0	32	2	0	0	0	1	0	0	0	0	0	0	0	0	0	0	Driessen / Macdonald, 97, 190
1374	Pak	Block Yps	1	1	3	1	4	32	2	0	0	0	1	0	0	0	0	0	0	0	0	0	0	Driessen / Macdonald, 97, 230
1366	Pak	Block Chi	1	1	3	1	4	32	2	0	0	0	1	0	0	0	0	0	0	0	0	0	0	Driessen / Macdonald, 97, 228
1367	Pak	Block Chi	1	1	3	1	4	32	2	0	0	0	1	0	0	0	0	0	0	0	0	0	0	Driessen / Macdonald, 97, 228
1370	Pak	Block Ksi	1	1	3	1	4	32	2	0	0	0	1	0	0	0	0	0	0	0	0	0	0	Driessen / Macdonald, 97, 230

Appendix V.2 Bronze objects. Sorted by object type / period / site. Total 1857

No	Site	Co	G	L	S	Pe	Da	Ot	C	TM	TK	TP	Cn	Le	Wi	He	Rd	We	Cu	Sn	As	Pb	Mu	Ref
1368	Pak	Block Chi	1	1	3	1	4	32	2	0	0	0	1	0	0	0	0	0	0	0	0	0	0	Driessen / Macdonald, 97, 228
1371	Pak	Block Ksi	1	1	3	1	4	32	2	0	0	0	1	0	0	0	0	0	0	0	0	0	0	Driessen / Macdonald, 97, 230
1372	Pak	Block Ksi	1	1	3	1	4	32	2	0	0	0	1	0	0	0	0	0	0	0	0	0	0	Driessen / Macdonald, 97, 230
1332	Psy	0	1	2	7	1	0	32	2	0	0	0	3	7.3	0	0	0	0	0	0	0	0	AshM	Boardman, 61, 16, no 53
1330	Psy	0	1	2	7	1	0	32	2	0	0	0	3	8.3	0	0	0	0	0	0	0	0	AshM	Boardman, 61, 16, no 51
1331	Psy	0	1	2	7	1	0	32	2	0	0	1	1	10	0	0	0	0	0	0	0	0	AshM	Boardman, 61, 16, no 53
1337	Psy	0	1	2	7	1	0	32	2	0	0	3	1	23	0	0	0	0	0	0	0	0	447	Boardman, 61, 17
1336	Psy	0	1	2	7	1	0	32	2	0	0	3	3	13	0	0	0	0	0	0	0	0	449	Boardman, 61, 17
1338	Psy	0	1	2	7	1	0	32	2	0	0	1	1	12	0	0	0	0	0	0	0	0	451	Boardman, 61, 17
1339	Psy	0	1	2	7	1	0	32	2	0	0	1	1	11	0	0	0	0	0	0	0	0	453	Boardman, 61, 17
1333	Psy	0	1	2	7	1	0	32	2	0	0	1	2	11	0	0	0	0	0	0	0	0	AshM	Boardman, 61, 16, no 54
1329	Psy	0	1	2	7	1	0	32	2	0	0	1	1	24	0	0	0	0	0	0	0	0	AshM	Boardman, 61, 16, no 51
1334	Psy	0	1	2	7	1	0	32	2	0	0	3	1	23	0	0	0	0	0	0	0	0	448	Boardman, 61, 17
1335	Psy	0	1	2	7	1	0	32	2	0	0	3	1	24	0	0	0	0	0	0	0	0	HerM	Boardman, 61,17
1470	Cha	0	4	1	6	2	6	32	2	0	0	2	2	0	0	0	0	0	0	0	0	0	0	Kanta, 80, 226; Andreadaki, 97, 505
1469	Cha	0	4	1	6	2	6	32	2	0	0	2	2	0	0	0	0	0	0	0	0	0	0	Kanta, 80, 226; Andreadaki, 97, 505
1350	Hal	Tomb 2	2	2	6	2	5	32	2	0	0	2	2	36	0	0	0	0	0	0	0	0	0	Hood, 56, 98, no 6
1351	Hal	Tomb 2	2	2	6	2	5	32	2	0	0	2	2	30	0	0	0	0	0	0	0	0	0	Hood, 56, 98, no 7
1356	Kno	UM	2	2	2	2	5	32	2	0	0	2	3	8.4	0	0	0	0	0	0	0	0	StraM	Driessen / Macdonald, 84, 71, no Eii 1
1349	Mas	Tomb XVIII	2	2	6	2	6	32	2	0	0	2	1	28	0	0	0	0	0	0	0	0	2142	Forsdyke, 26-27, 282, no XVIII.2
1352	NeH	Tomb III	2	2	6	2	6	32	2	0	0	1	1	29	0	0	0	0	0	0	0	0	0	Hood / d.Jong, 52, 269, no III.13
1459	Pig	0	4	2	6	2	6	32	2	0	0	0	1	0	0	0	0	0	0	0	0	0	0	Kanta, 80, 213
1460	Pig	0	4	2	6	2	6	32	2	0	0	0	1	0	0	0	0	0	0	0	0	0	0	Kanta, 80, 213
1461	Pig	0	4	2	6	2	6	32	2	0	0	0	1	0	0	0	0	0	0	0	0	0	0	Kanta, 80, 213
1353	Sel	Tomb 4/II	2	2	6	2	6	32	2	0	0	2	2	21	0	0	0	0	0	0	0	0	0	Catling / Catling, 74, 229, no 3
1348	ZaP	Tomb 86	2	2	6	2	6	32	2	0	0	3	1	23	0	0	0	0	0	0	0	0	0	Evans, 06, 82, no 86a
1347	ZaP	Tomb 62	2	2	6	2	6	32	2	0	0	1	1	19	0	0	0	0	0	0	0	0	0	Evans, 06, 68, no 62c
1845	Gal	0	3	2	6	3	7	32	2	0	0	2	1	34	0	0	0	0	0	0	0	0	0	Driessen / Macdonald, 84, 71, no Eii 4
1443	HaG	0	1	2	6	3	0	32	2	0	0	0	1	0	0	0	0	0	0	0	0	0	0	Kanta, 80, 164, 178
1441	Paa	Pithos Bur	1	1	6	3	7	32	2	0	0	0	1	0	0	0	0	0	0	0	0	0	0	Kanta, 80, 143
1458	Pak	Ellinika	1	1	6	3	0	32	2	0	0	0	1	0	0	0	0	0	0	0	0	0	0	Kanta, 80, 191
1453	Phk	0	1	2	6	3	7	32	2	0	0	0	1	0	0	0	0	0	0	0	0	0	0	Kanta, 80, 164, 183
1433	Pom	0	3	2	6	3	8	32	2	0	0	0	1	0	0	0	0	0	0	0	0	0	0	Kanta, 80, 95
1342	Psy	0	1	2	7	3	0	32	2	0	0	1	3	8.1	0	0	0	0	0	0	0	0	HerM	Boardman, 61,17
1341	Psy	0	1	2	7	3	8	32	2	0	0	2	1	10	0	0	0	0	0	0	0	0	327	Boardman, 61,17
1340	Psy	0	1	2	7	3	8	32	2	0	0	2	1	9.5	0	0	0	0	0	0	0	0	326	Boardman, 61,17
1475	Sam	0	4	2	5	3	8	32	2	0	0	2	1	0	0	0	0	0	0	0	0	0	0	Kanta, 80, 236
1345	Sit	0	1	0	0	3	0	32	2	0	0	2	1	25	0	0	0	0	0	0	0	0	AshM	Catling, 68, 89, 90, no 4
1344	Sit	0	1	0	0	3	0	32	2	0	0	2	2	28	0	0	0	0	0	0	0	0	AshM	Catling, 68, 89, 90, no 3
1346	Sit	0	1	0	0	3	0	32	2	0	0	2	1	34	0	0	0	0	0	0	0	0	AshM	Catling, 68, 89, 92, no 5; Baboula / Northover, 99, 147

Appendix V.2 Bronze objects. Sorted by object type / period / site. Total 1857

No	Site	Co	G	L	S	Pe	Da	Ot	C	TM	TK	TP	Cn	Le	Wi	He	Rd	We	Cu	Sn	As	Pb	Mu	Ref
1408	Stm	Tomb C	2	2	6	3	7	32	2	0	0	0	0	0	0	0	0	0	0	0	0	0	0	Kanta, 80, 53
1414	Stm	Tomb E	2	2	6	3	8	32	2	0	0	0	0	0	0	0	0	0	0	0	0	0	0	Kanta, 80, 54
1290	0	0	0	0	0	0	0	33	2	0	0	0	2	19	0	0	0	0	0	0	0	0	AshM	Catling, 68, 94, no 8
1289	0	0	0	0	0	0	0	33	2	0	0	0	2	30	0	0	0	0	0	0	0	0	AshM	Catling, 68, 92, no 7
689	Che	0	2	1	0	0	0	33	2	0	0	0	2	32	0	0	0	0	0	0	0	0	GiaC	Avila, 83, 26, no 52
1098	Cre	0	0	0	0	0	0	33	2	0	0	0	1	22	0	0	0	0	0	0	0	0	0	Höckmann, 80, 131, no A 6
1096	Cre	0	0	0	0	0	0	33	2	0	0	0	1	23	0	0	0	0	0	0	0	0	0	Höckmann, 80, 138, no E 6
692	Cre	0	0	0	0	0	0	33	2	0	0	0	0	0	0	0	0	0	0	0	0	0	GiaC	Avila, 83, 12, no 21
1510	Psy	0	1	2	7	0	0	33	2	0	0	0	1	25	0	0	0	0	90.2	9.4	0.3	0	AshM	Boardman, 61, 28, no 98, 160
1517	Psy	0	1	2	7	0	0	33	2	0	0	0	1	8.3	0	0	0	0	0	0	0	0	AshM	Boardman, 61, 28 ,no 105
1523	Psy	0	1	2	7	0	0	33	2	0	0	0	1	11	0	0	0	0	0	0	0	0	AshM	Boardman, 61, 28, no 111
1513	Psy	0	1	2	7	0	0	33	2	0	0	0	1	9	0	0	0	0	0	0	0	0	AshM	Boardman, 61, 28, no 101
1514	Psy	0	1	2	7	0	0	33	2	0	0	0	1	8.7	0	0	0	0	0	0	0	0	AshM	Boardman, 61, 28, no 102
1519	Psy	0	1	2	7	0	0	33	2	0	0	0	1	5.5	0	0	0	0	0	0	0	0	AshM	Boardman, 61, 28, no 107
1511	Psy	0	1	2	7	0	0	33	2	0	0	0	1	20	0	0	0	0	95.2	3.4	0.6	0.1	AshM	Boardman, 61, 28, no 99, 160
1509	Psy	0	1	2	7	0	0	33	2	0	0	0	1	29	0	0	0	0	88.2	7.9	1.7	0.1	AshM	Boardman, 61, 28, no 97, 160
1522	Psy	0	1	2	7	0	0	33	2	0	0	0	1	12	0	0	0	0	0	0	0	0	AshM	Boardman, 61, 28, no 110
1520	Psy	0	1	2	7	0	0	33	2	0	0	0	1	3.6	0	0	0	0	0	0	0	0	AshM	Boardman, 61, 28, no 108
1515	Psy	0	1	2	7	0	0	33	2	0	0	0	1	7.5	0	0	0	0	0	0	0	0	AshM	Boardman, 61, 28, no 103
1518	Psy	0	1	2	7	0	0	33	2	0	0	0	1	8.7	0	0	0	0	0	0	0	0	AshM	Boardman, 61, 28 ,no 106
1524	Psy	0	1	2	7	0	0	33	2	0	0	0	1	8.7	0	0	0	0	99	0.1	0	0.8	AshM	Boardman, 61, 28, no 112, 160
1521	Psy	0	1	2	7	0	0	33	2	0	0	0	1	5.4	0	0	0	0	0	0	0	0	AshM	Boardman, 61, 28, no 109
1516	Psy	0	1	2	7	0	0	33	2	0	0	0	1	4.5	0	0	0	0	0	0	0	0	AshM	Boardman, 61, 28, no 104
1512	Psy	0	1	2	7	0	0	33	2	0	0	0	2	9	0	0	0	0	93.1	6.6	0	0.1	AshM	Boardman, 61, 28, no 100, 160
683	Sit	0	1	0	0	0	0	33	2	0	0	0	2	42	0	0	0	0	0	0	0	0	AshM	Avila, 83, 23, no 47
690	Gou	A 8	1	1	3	1	4	33	2	0	0	0	1	30	0	0	0	0	0	0	0	0	558	Avila, 83, 26, no 53
1646	Gou	A 8	1	1	3	1	4	33	2	0	0	0	1	30	0	0	0	0	0	0	0	0	558	Boyd Hawes, 08, 34, no 48
1798	HaT	Ro 31-32	3	2	4	1	3	33	2	0	0	0	1	0	0	0	0	0	0	0	0	0	0	Driessen / Macdonald, 97, 202
1797	HaT	Ro 31-32	3	2	4	1	3	33	2	0	0	0	1	0	0	0	0	0	0	0	0	0	0	Driessen / Macdonald, 97, 202
687	Iso	Iso Dep	2	2	6	1	0	33	2	0	0	0	3	33	0	0	0	0	0	0	0	0	0	Evans, 14, 4, fig 6
1099	Iuk	0	2	2	8	1	2	33	2	0	0	0	1	0	0	0	0	0	0	0	0	0	0	Höckmann, 80, 131, no A 8
1091	Mal	House Z	2	1	2	1	4	33	2	0	0	0	1	24	0	0	0	0	0	0	0	0	0	Höckmann, 80, 139, no F 7
1086	Moc	Tomb XX	1	1	6	1	0	33	2	0	0	0	1	33	0	0	0	0	0	0	0	0	0	Höckmann, 80, 133, no C15
1087	Moc	Tomb XX	1	1	6	1	0	33	2	0	0	0	1	28	0	0	0	0	0	0	0	0	0	Höckmann, 80, 139, no F6
1088	Moc	Tomb XX	1	1	6	1	0	33	2	0	0	0	1	28	0	0	0	0	0	0	0	0	0	Höckmann, 80, 141, no G3
1092	Pse	0	1	1	3	1	0	33	2	0	0	0	1	26	0	0	0	0	0	0	0	0	1588	Höckmann, 80, 141, no F 47
1080	Akr	0	2	2	6	2	5	33	2	0	0	0	1	18	0	0	0	0	0	0	0	0	0	Höckmann, 80, 133, no C13
1081	Akr	0	2	2	6	2	5	33	2	0	0	0	1	30	0	0	0	0	0	0	0	0	0	Höckmann, 80, 134, no D12
691	Arh	Tholos B	2	2	6	2	5	33	2	0	0	0	1	20	0	0	0	0	0	0	0	0	HerM	Avila, 83, 134, no 871
696	Arm	Tomb 200	4	2	6	2	6	33	2	0	0	0	3	4.8	0	0	0	0	0	0	0	0	0	Papadopoulou, 97, 334

Appendix V.2 Bronze objects. Sorted by object type / period / site. Total 1857

No	Site	Co	G	L	S	Pe	Da	Ot	C	TM	TK	TP	Cn	Le	Wi	He	Rd	We	Cu	Sn	As	Pb	Mu	Ref
688	Axo	0	4	2	0	2	5	33	2	0	0	0	1	49	0	0	0	0	0	0	0	0	GiaC	Avila, 83, 18, no 35
1472	Cha	0	4	1	6	2	6	33	2	0	0	0	1	0	0	0	0	0	0	0	0	0	0	Kanta, 80, 226
1471	Cha	0	4	1	6	2	6	33	2	0	0	0	1	0	0	0	0	0	0	0	0	0	0	Kanta, 80, 226
1095	Cre	0	0	0	0	2	5	33	2	0	0	0	1	38	0	0	0	0	0	0	0	0	GiaC	Höckmann, 80, 137, no D 53
671	Hal	Tomb 2	2	2	6	2	5	33	2	0	0	0	2	33	0	0	0	0	0	0	0	0	HerM	Avila, 83, 31, no 67
667	Hal	Tomb 1	2	2	6	2	5	33	2	0	0	0	1	43	0	0	0	0	0	0	0	0	HerM	Avila, 83, 25, no 51
673	Hal	Tomb 2	2	2	6	2	5	33	2	0	0	0	1	36	0	0	0	0	0	0	0	0	HerM	Avila, 83, 134, no 876
670	Hal	Tomb 2	2	2	6	2	5	33	2	0	0	0	1	41	0	0	0	0	0	0	0	0	HerM	Avila, 83, 26, no 55
674	Hal	Tomb 2	2	2	6	2	5	33	2	0	0	0	1	24	0	0	0	0	0	0	0	0	HerM	Avila, 83, 134, no 878
668	Hal	Tomb 1	2	2	6	2	5	33	2	0	0	0	1	38	0	0	0	0	0	0	0	0	HerM	Avila, 83, 132, no 856
669	Hal	Tomb 1	2	2	6	2	5	33	2	0	0	0	1	24	0	0	0	0	0	0	0	0	HerM	Avila, 83, 134, no 877
672	Hal	Tomb 2	2	2	6	2	5	33	2	0	0	0	1	50	0	0	0	0	0	0	0	0	HerM	Avila, 83, 133, no 861
675	Hal	Tomb 2	2	2	6	2	5	33	2	0	0	0	3	18	0	0	0	0	0	0	0	0	0	Hood, 56, 98, no 13
1089	HaT	0	3	20	0	2	5	33	2	0	0	0	1	29	0	0	0	0	0	0	0	0	1244	Höckmann, 80, 137, no D 50
686	Iso	Tomb 3	2	2	6	2	6	33	2	0	0	0	1	47	0	0	0	0	0	0	0	0	HerM	Avila, 83, 133, no 867
1477	Kno	UM H	2	2	2	2	5	33	2	0	0	0	2	31	0	0	0	0	83	17	0	0	0	Popham, 84, 26, no H 62; Catling / Catling, 84, 212
1852	Kno	UM M	2	2	2	2	5	33	2	0	0	0	1	20	3.1	0	0	0	89	9	2	0	0	Popham, 84, 61, no M 90
1478	Kno	UM H	2	2	2	2	5	33	2	0	0	0	2	26	0	0	0	0	89	11	0	0	0	Popham, 84, 26, no H 63; Catling / Catling, 84, 212
693	Mas	Tomb XVIII	2	2	6	2	6	33	2	0	0	0	1	20	0	0	0	0	0	0	0	0	0	Forsdyke, 26-27, 282, no 3
677	NeH	Tomb III	2	2	6	2	5	33	2	0	0	0	1	47	0	0	0	0	0	0	0	0	HerM	Avila, 83, 133, no 865
1082	NeH	Tomb I	2	2	6	2	5	33	2	0	0	0	1	25	0	0	0	0	0	0	0	0	0	Höckmann, 80, 134, no D13
678	NeH	Tomb V	2	2	6	2	5	33	2	0	0	0	1	49	0	0	0	0	0	0	0	0	HerM	Avila, 83, 133, no 866
676	NeH	Tomb II	2	2	6	2	6	33	2	0	0	0	1	56	0	0	0	0	0	0	0	0	HerM	Avila, 83, 22, no 43
685	Sel	Tomb 4/I	2	2	6	2	6	33	2	0	0	0	1	27	0	0	0	0	0	0	0	0	2974	Popham / Catling 74, 229, no 5
682	Sel	Tomb 4/I	2	2	6	2	6	33	2	0	0	0	1	35	0	0	0	0	0	0	0	0	2972	Avila, 83, 23, no 46
694	ZaP	Tomb 14	2	2	6	2	6	33	2	0	0	0	1	25	0	0	0	0	0	0	0	0	0	Evans, 06, 43, no 14q
695	ZaP	Tomb 55	2	2	6	2	6	33	2	0	0	0	1	24	0	0	0	0	0	0	0	0	0	Evans, 06, 66, no 55c
679	ZaP	Tomb 75	2	2	6	2	6	33	2	0	0	0	1	27	0	0	0	0	0	0	0	0	HerM	Avila, 83, 21, no 41
1844	Arh	Bur.Buil. 3	2	2	6	3	0	33	2	0	0	0	1	0	0	0	0	0	0	0	0	0	HerM	Sakellarakis / Sakellarakis,.97, 598
1097	Cre	0	0	0	0	3	0	33	2	0	0	0	1	22	0	0	0	0	0	0	0	0	GiaC	Höckmann, 80, 146, no H 35
1093	Cre	0	0	0	0	3	0	33	2	0	0	0	1	42	0	0	0	0	0	0	0	0	AshM	Höckmann, 80, 135, no D 25
1094	Cre	0	0	0	0	3	0	33	2	0	0	0	1	32	0	0	0	0	0	0	0	0	0	Höckmann, 80, 137, no D 46
1357	Gal	0	3	2	6	3	7	33	2	0	0	0	2	0	0	0	0	0	0	0	0	0	0	Driessen / Macdonald, 84, 71, no Eii 4
1100	Gal	0	0	0	0	3	7	33	2	0	0	0	1	28	0	0	0	0	0	0	0	0	0	Höckmann, 80, 134, no D 5
1090	Kno	Ellinika	2	2	0	3	0	33	2	0	0	0	1	40	0	0	0	0	0	0	0	0	1783	Höckmann, 80, 137, no D 51
1836	Mol	Tomb B	1	1	6	3	9	33	2	0	0	0	1	9.5	0	0	0	0	0	0	0	0	0	Höckmann, 80, 149, no K23
1837	Mol	Tomb B	1	1	6	3	9	33	2	0	0	0	1	19	0	0	0	0	0	0	0	0	0	Höckmann, 80, 150, no K 27
1835	Mol	Tomb A	1	1	6	3	9	33	2	0	0	0	2	21	0	0	0	0	0	0	0	0	0	Höckmann, 80, 144, no H12
1429	NeC	0	3	2	6	3	0	33	2	0	0	0	1	0	0	0	0	0	0	0	0	0	0	Kanta, 80, 83
1428	NeC	0	3	2	6	3	0	33	2	0	0	0	1	0	0	0	0	0	0	0	0	0	0	Kanta, 80, 83

Appendix V.2 Bronze objects. Sorted by object type / period / site. Total 1857

No	Site	Co	G	L	S	Pe	Da	Ot	C	TM	TK	TP	Cn	Le	Wi	He	Rd	We	Cu	Sn	As	Pb	Mu	Ref
1455	Phk	0	1	2	6	3	7	33	2	0	0	0	1	0	0	0	0	0	0	0	0	0	0	Kanta, 80, 164, 183
1454	Phk	0	1	2	6	3	7	33	2	0	0	0	1	0	0	0	0	0	0	0	0	0	0	Kanta, 80, 164, 183
1084	Stm	Tomb E	2	2	6	3	8	33	2	0	0	0	1	18	0	0	0	0	0	0	0	0	0	Höckmann, 80, 141, no F26
1085	Stm	Tomb B	2	2	6	3	8	33	2	0	0	0	1	17	0	0	0	0	0	0	0	0	0	Höckmann, 80, 148, no K10
1409	Stm	Tomb D	2	2	6	3	7	33	2	0	0	0	1	0	0	0	0	0	0	0	0	0	0	Kanta, 80, 53
1083	Stm	Tomb B	2	2	6	3	8	33	2	0	0	0	1	17	0	0	0	0	0	0	0	0	0	Höckmann, 80, 138, no E5
684	Tyl	Crem. To	4	2	6	3	0	33	2	0	0	0	3	23	0	0	0	0	0	0	0	0	2213	Avila, 83, 23, no 43A
1437	Vrs	0	1	2	6	3	7	33	2	0	0	0	1	0	0	0	0	0	0	0	0	0	0	Kanta, 80, 141
680	ZaP	Tomb 36	2	2	6	3	7	33	2	0	0	0	1	34	0	0	0	0	0	0	0	0	HerM	Avila, 83, 133, no 869
681	ZaP	Tomb 36	2	2	6	3	7	33	2	0	0	0	1	26	0	0	0	0	0	0	0	0	HerM	Avila, 83, 133, no 870
1529	Psy	0	1	2	7	0	0	34	2	0	0	0	1	7.6	0	0	0	0	0	0	0	0	AshM	Boardman, 61, 30, no 118
1532	Psy	0	1	2	7	0	0	34	2	0	0	0	1	2.9	0	0	0	0	0	0	0	0	AshM	Boardman, 61, 30, no 123
1531	Psy	0	1	2	7	0	0	34	2	0	0	0	1	4.7	0	0	0	0	0	0	0	0	AshM	Boardman, 61, 30, no 122
729	Mas	Tomb V	2	2	6	1	4	34	2	0	0	0	1	3.5	0	0	0	0	0	0	0	0	0	Forsdyke, 26-27, 257, no 1
728	Mas	Tomb V	2	2	6	1	4	34	2	0	0	0	1	3.5	0	0	0	0	0	0	0	0	0	Forsdyke, 26-27, 257, no 1
1824	Tyl	House A	4	2	4	1	4	34	2	0	0	0	1	0	0	0	0	0	0	0	0	0	0	Driessen / Macdonald, 97, 129
1825	Tyl	House A	4	2	4	1	4	34	2	0	0	0	1	0	0	0	0	0	0	0	0	0	0	Driessen / Macdonald, 97, 129
706	Hal	Tomb 2	2	2	6	2	5	34	2	0	0	0	1	9	0	0	0	0	0	0	0	0	HerM	Avila, 83, 110, no 756
707	Hal	Tomb 2	2	2	6	2	5	34	2	0	0	0	1	8.5	0	0	0	0	0	0	0	0	HerM	Avila, 83, 111, no 758B
710	Hal	Tomb 2	2	2	6	2	5	34	2	0	0	0	1	9.1	0	0	0	0	0	0	0	0	HerM	Avila, 83, 113, no 770G
708	Hal	Tomb 2	2	2	6	2	5	34	2	0	0	0	1	8.6	0	0	0	0	0	0	0	0	HerM	Avila, 83, 113, no 770E
709	Hal	Tomb 2	2	2	6	2	5	34	2	0	0	0	1	8.9	0	0	0	0	0	0	0	0	HerM	Avila, 83, 113, no 770F
747	Iso	Tomb 2	2	2	6	2	5	34	2	0	0	0	1	5.1	0	0	0	0	0	0	0	0	0	Evans, 14, 58, no 2f, fig 54
756	Iso	Tomb 2	2	2	6	2	5	34	2	0	0	0	2	5.2	0	0	0	0	0	0	0	0	0	Evans, 14, 58, no 2f, fig 54
755	Iso	Tomb 2	2	2	6	2	5	34	2	0	0	0	1	5.2	0	0	0	0	0	0	0	0	0	Evans, 14, 58, no 2f, fig 54
741	Iso	Tomb 2	2	2	6	2	5	34	2	0	0	0	2	4.9	0	0	0	0	0	0	0	0	0	Evans, 14, 58, no 2f, fig 54
740	Iso	Tomb 2	2	2	6	2	5	34	2	0	0	0	1	4.9	0	0	0	0	0	0	0	0	0	Evans, 14, 58, no 2f, fig 54
742	Iso	Tomb 2	2	2	6	2	5	34	2	0	0	0	1	5	0	0	0	0	0	0	0	0	0	Evans, 14, 58, no 2f, fig 54
745	Iso	Tomb 2	2	2	6	2	5	34	2	0	0	0	1	5	0	0	0	0	0	0	0	0	0	Evans, 14, 58, no 2f, fig 54
750	Iso	Tomb 2	2	2	6	2	5	34	2	0	0	0	1	5.1	0	0	0	0	0	0	0	0	0	Evans, 14, 58, no 2f, fig 54
748	Iso	Tomb 2	2	2	6	2	5	34	2	0	0	0	1	5.1	0	0	0	0	0	0	0	0	0	Evans, 14, 58, no 2f, fig 54
751	Iso	Tomb 2	2	2	6	2	5	34	2	0	0	0	2	5.1	0	0	0	0	0	0	0	0	0	Evans, 14, 58, no 2f, fig 54
733	Iso	Tomb1A	2	2	6	2	6	34	2	0	0	0	1	0	0	0	0	0	0	0	0	0	0	Evans, 14, 2, 6
732	Iso	Tomb1A	2	2	6	2	6	34	2	0	0	0	1	5.5	0	0	0	0	0	0	0	0	0	Evans, 14, 2, 6, fig 10
752	Iso	Tomb 2	2	2	6	2	5	34	2	0	0	0	1	5.2	0	0	0	0	0	0	0	0	0	Evans, 14, 58, no 2f, fig 54
734	Iso	Tomb1A	2	2	6	2	6	34	2	0	0	0	1	0	0	0	0	0	0	0	0	0	0	Evans, 14, 2, 6
743	Iso	Tomb 2	2	2	6	2	5	34	2	0	0	0	1	5	0	0	0	0	0	0	0	0	0	Evans, 14, 58, no 2f. fig 54
735	Iso	Tomb1A	2	2	6	2	6	34	2	0	0	0	1	0	0	0	0	0	0	0	0	0	0	Evans, 14, 2, 6
749	Iso	Tomb 2	2	2	6	2	5	34	2	0	0	0	1	5.1	0	0	0	0	0	0	0	0	0	Evans, 14, 58, no 2f, fig 54
731	Iso	Tomb1A	2	2	6	2	6	34	2	0	0	0	1	7.5	0	0	0	0	0	0	0	0	0	Evans, 14, 2, 6, fig 10

Appendix V.2 Bronze objects. Sorted by object type / period / site. Total 1857

No	Site	Co	G	L	S	Pe	Da	Ot	C	TM	TK	TP	Cn	Le	Wi	He	Rd	We	Cu	Sn	As	Pb	Mu	Ref
736	Iso	Tomb 3	2	2	6	2	6	34	2	0	0	0	1	6.7	0	0	0	0	0	0	0	0	0	Evans, 14, 15, 3c
737	Iso	Tomb 2	2	2	6	2	5	34	2	0	0	0	1	4.9	0	0	0	0	0	0	0	0	0	Evans, 14, 58, no 2f, fig 54
730	Iso	Tomb1A	2	2	6	2	6	34	2	0	0	0	1	8	0	0	0	0	0	0	0	0	0	Evans, 14, 2, 6, fig 10
744	Iso	Tomb 2	2	2	6	2	5	34	2	0	0	0	1	5	0	0	0	0	0	0	0	0	0	Evans, 14, 58, no 2f, fig 54
739	Iso	Tomb 2	2	2	6	2	5	34	2	0	0	0	1	4.9	0	0	0	0	0	0	0	0	0	Evans, 14, 58, no 2f, fig 54
738	Iso	Tomb 2	2	2	6	2	5	34	2	0	0	0	1	4.9	0	0	0	0	0	0	0	0	0	Evans, 14, 58, no 2f, fig 54
746	Iso	Tomb 2	2	2	6	2	5	34	2	0	0	0	2	5	0	0	0	0	0	0	0	0	0	Evans, 14, 58, no 2f, fig 54
753	Iso	Tomb 2	2	2	6	2	5	34	2	0	0	0	1	5.2	0	0	0	0	0	0	0	0	0	Evans, 14, 58, no 2f, fig 54
754	Iso	Tomb 2	2	2	6	2	5	34	2	0	0	0	1	5.2	0	0	0	0	0	0	0	0	0	Evans, 14, 58, no 2f, fig 54
712	Kno	Armoury	2	2	1	2	5	34	2	0	0	0	1	4.3	0	0	0	0	0	0	0	0	HerM	Avila, 83, 105, no 724
711	Kno	Armoury	2	2	1	2	5	34	2	0	0	0	1	4.3	0	0	0	0	0	0	0	0	HerM	Avila, 83, 94, no 597
1479	Kno	UM H	2	2	2	2	5	34	2	0	0	0	1	4.8	0	0	0	0	100	0	0	0	0	Popham, 84, 37, no H 193; Catling / Catling, 84, 212
703	NeH	Tomb III	2	2	6	2	5	34	2	0	0	0	1	9.2	0	0	0	0	0	0	0	0	HerM	Avila, 83, 113, no 770D
701	NeH	Tomb III	2	2	6	2	5	34	2	0	0	0	1	6.5	0	0	0	0	0	0	0	0	HerM	Avila, 83, 110, no 757
700	NeH	Tomb III	2	2	6	2	5	34	2	0	0	0	1	3.1	0	0	0	0	0	0	0	0	HerM	Avila, 83, 105, no 723B
702	NeH	Tomb III	2	2	6	2	5	34	2	0	0	0	1	5.2	0	0	0	0	0	0	0	0	HerM	Avila, 83, 110, no 758
705	NeH	Tomb III	2	2	6	2	5	34	2	0	0	0	1	7.2	0	0	0	0	0	0	0	0	HerM	Avila, 83, 113, no 770N
704	NeH	Tomb III	2	2	6	2	5	34	2	0	0	0	1	9	0	0	0	0	0	0	0	0	HerM	Avila, 83, 113, no 770M
1528	Psy	0	1	2	7	2	5	34	2	0	0	0	1	7	0	0	0	0	0	0	0	0	AshM	Boardman, 61, 30, no 117
1527	Psy	0	1	2	7	2	5	34	2	0	0	0	1	6.5	0	0	0	0	0	0	0	0	AshM	Boardman, 61, 30, no 116
699	Sel	Tomb 3	2	2	6	2	6	34	2	0	0	0	2	3.2	0	0	0	0	0	0	0	0	HerM	Avila, 83, 92, no 579
698	Sel	Tomb 3	2	2	6	2	6	34	2	0	0	0	2	4.7	0	0	0	0	0	0	0	0	HerM	Avila, 83, 90, no 551
697	Sel	Tomb 3	2	2	6	2	6	34	2	0	0	0	2	3.7	0	0	0	0	0	0	0	0	HerM	Avila, 83, 90, no 550
727	ZaP	Tomb 10	2	2	6	2	6	34	2	0	0	0	1	0	0	0	0	0	0	0	0	0	0	Evans, 06, 32, fig 27, 28
716	ZaP	Tomb 10	2	2	6	2	6	34	2	0	0	0	1	0	0	0	0	0	0	0	0	0	0	Evans, 06, 32, fig 27, 28
713	ZaP	Tomb 10	2	2	6	2	6	34	2	0	0	0	1	0	0	0	0	0	0	0	0	0	0	Evans, 06, 32, fig 27, 28
719	ZaP	Tomb 10	2	2	6	2	6	34	2	0	0	0	1	0	0	0	0	0	0	0	0	0	0	Evans, 06, 32, fig 27, 28
717	ZaP	Tomb 10	2	2	6	2	6	34	2	0	0	0	1	0	0	0	0	0	0	0	0	0	0	Evans, 06, 32, fig 27, 28
714	ZaP	Tomb 10	2	2	6	2	6	34	2	0	0	0	1	0	0	0	0	0	0	0	0	0	0	Evans, 06, 32, fig 27, 28
725	ZaP	Tomb 10	2	2	6	2	6	34	2	0	0	0	1	0	0	0	0	0	0	0	0	0	0	Evans, 06, 32, fig 27, 28
726	ZaP	Tomb 10	2	2	6	2	6	34	2	0	0	0	1	0	0	0	0	0	0	0	0	0	0	Evans, 06, 32, fig 27, 28
722	ZaP	Tomb 10	2	2	6	2	6	34	2	0	0	0	1	0	0	0	0	0	0	0	0	0	0	Evans, 06, 32, fig 27, 28
724	ZaP	Tomb 10	2	2	6	2	6	34	2	0	0	0	1	0	0	0	0	0	0	0	0	0	0	Evans, 06, 32, fig 27, 28
721	ZaP	Tomb 10	2	2	6	2	6	34	2	0	0	0	1	0	0	0	0	0	0	0	0	0	0	Evans, 06, 32, fig 27, 28
723	ZaP	Tomb 10	2	2	6	2	6	34	2	0	0	0	1	0	0	0	0	0	0	0	0	0	0	Evans, 06, 32, fig 27, 28
715	ZaP	Tomb 10	2	2	6	2	6	34	2	0	0	0	1	0	0	0	0	0	0	0	0	0	0	Evans, 06, 32, fig 27, 28
718	ZaP	Tomb 10	2	2	6	2	6	34	2	0	0	0	1	0	0	0	0	0	0	0	0	0	0	Evans, 06, 32, fig 27, 28
720	ZaP	Tomb 10	2	2	6	2	6	34	2	0	0	0	1	0	0	0	0	0	0	0	0	0	0	Evans, 06, 32, fig 27, 28
1709	Cha	0	4	1	3	3	8	34	2	0	0	0	1	0	0	0	0	0	94	5.8	0.5	0	ChaM	Stos-Gale.et.al, 00, 213, Tabl 2, 214, Tabl 3, no 23
1526	Psy	0	1	2	7	3	0	34	2	0	0	0	3	2.6	0	0	0	0	0	0	0	0	AshM	Boardman, 61, 29, 30, no 115

Appendix V.2 Bronze objects. Sorted by object type / period / site. Total 1857

No	Site	Co	G	L	S	Pe	Da	Ot	C	TM	TK	TP	Cn	Le	Wi	He	Rd	We	Cu	Sn	As	Pb	Mu	Ref
1530	Psy	0	1	2	7	3	0	34	2	0	0	0	0	5.4	0	0	0	0	0	0	0	0	AshM	Boardman, 61, 30, no 121
1525	Psy	0	1	2	7	3	0	34	2	0	0	0	0	3.5	0	0	0	0	0	0	0	0	AshM	Boardman, 61, 29, 30, no 114
1283	NeH	Tomb V	2	2	6	2	6	36	2	0	0	0	3	0	21	24	0	1123	0	0	0	0	0	Hood / d.Jong, 52, 256, 275, no 8
1826	Apd	0	4	2	5	1	4	45	3	0	0	0	1	0	0	0	0	0	0	0	0	0	0	Driessen / Macdonald, 97, 132
1827	Apd	0	4	2	5	1	4	45	3	0	0	0	1	0	0	0	0	0	0	0	0	0	0	Driessen / Macdonald, 97, 132
1703	Gou	Hill House	1	1	3	1	4	45	3	0	0	0	1	25	15	0	0	0	0	0	0	0	556	Boyd Hawes, 08, 48, no 23
1702	Gou	C 31	1	1	3	1	4	45	3	0	0	0	1	31	15	0	0	0	0	0	0	0	964	Boyd Hawes, 08, 48, no 22
1780	HaT	0	3	2	4	1	4	45	3	0	0	0	1	0	0	0	0	0	0	0	0	0	736	Halbherr et al, 80, 68
1778	HaT	Room 57a	3	2	4	1	0	45	3	0	0	0	1	0	0	0	0	0	0	0	0	0	0	Halbherr et al, 80, 123
1781	HaT	0	3	2	4	1	4	45	3	0	0	0	1	0	0	0	0	0	0	0	0	0	739	Halbherr et al, 80, 68
1779	HaT	Casa Est	3	2	4	1	3	45	3	0	0	0	1	0	0	0	0	0	0	0	0	0	1159	Halbherr et al, 80, 287
1800	Kno	Well SoP	2	2	2	1	0	45	3	0	0	0	3	0	0	0	0	0	0	0	0	0	0	Evans, 14, 53-54
1799	Kno	Well SoP	2	2	2	1	0	45	3	0	0	0	1	19	0	0	0	0	0	0	0	0	0	Evans, 14, 53-54
1744	Moc	House C3	1	1	3	1	4	45	3	0	0	0	3	0	0	0	0	0	0	0	0	0	0	Soles / Davaras, 96, 194, no CA 54
1506	NiC	Room 7	2	1	4	1	0	45	3	0	0	0	2	100	60	0	0	0	0	0	0	0	0	Georgiou, 79, 43, no 4
1503	NiC	Room 7	2	1	4	1	0	45	3	0	0	0	1	118	24	0	0	0	0	0	0	0	0	Georgiou, 79, 43, no 1
1504	NiC	Room 7	2	1	4	1	0	45	3	0	0	0	1	93	53	0	0	0	0	0	0	0	0	Georgiou, 79, 43, no 2
1505	NiC	Room 7	2	1	4	1	0	45	3	0	0	0	1	93	53	0	0	0	0	0	0	0	0	Georgiou, 79, 43, no 3
1379	Pak	Kour.	1	1	3	1	4	45	3	0	0	0	2	25	0	0	0	0	0	0	0	0	0	Dawkins / Tod, 02-03, 333
1378	Pak	Kour.	1	1	3	1	4	45	3	0	0	0	2	24	0	0	0	0	0	0	0	0	0	Dawkins / Tod, 02-03, 333
1561	Psy	0	1	2	7	1	0	45	3	0	0	0	1	0	0	0	0	0	0	0	0	0	484	Boardman, 61, 42, no 5
1556	Psy	0	1	2	7	1	0	45	3	0	0	0	1	0	0	0	0	0	0	0	0	0	484	Boardman, 61, 42, no 2
1559	Psy	0	1	2	7	1	0	45	3	0	0	0	1	0	0	0	0	0	0	0	0	0	484	Boardman, 61, 42, no 3
1554	Psy	0	1	2	7	1	0	45	3	0	0	0	1	0	0	0	0	0	0	0	0	0	484	Boardman, 61, 42, no 2
1553	Psy	0	1	2	7	1	0	45	3	0	0	0	1	0	0	0	0	0	0	0	0	0	484	Boardman, 61, 42, no 2
1564	Psy	0	1	2	7	1	0	45	3	0	0	0	1	0	0	0	0	0	0	0	0	0	484	Boardman, 61, 42, no 8
1563	Psy	0	1	2	7	1	0	45	3	0	0	0	1	0	0	0	0	0	0	0	0	0	484	Boardman, 61, 42, no 8
1558	Psy	0	1	2	7	1	0	45	3	0	0	0	1	0	0	0	0	0	0	0	0	0	483	Boardman, 61, 42, no 3
1557	Psy	0	1	2	7	1	0	45	3	0	0	0	1	0	0	0	0	0	0	0	0	0	482	Boardman, 61, 42, no 3
1562	Psy	0	1	2	7	1	0	45	3	0	0	0	1	0	0	0	0	0	0	0	0	0	484	Boardman, 61, 42, no 8
1560	Psy	0	1	2	7	1	0	45	3	0	0	0	1	0	0	0	0	0	0	0	0	0	480	Boardman, 61, 42, no 4
1552	Psy	0	1	2	7	1	0	45	3	0	0	0	1	25	11	0	0	0	0	0	0	0	481	Boardman, 61, 42, no 2
1543	Psy	0	1	2	7	1	0	45	3	0	0	0	1	4.8	0	0	0	0	99.5	0.1	0.2	0.1	AshM	Boardman, 61, 44, 45, no 203, 160
1545	Psy	0	1	2	7	1	0	45	3	0	0	0	1	12	0	0	0	0	0	0	0	0	AshM	Boardman, 61, 44, 45, no 205
1548	Psy	0	1	2	7	1	0	45	3	0	0	0	1	7.1	0	0	0	0	0	0	0	0	AshM	Boardman, 61, 44, 45, no 208
1547	Psy	0	1	2	7	1	0	45	3	0	0	0	1	7.4	0	0	0	0	0	0	0	0	AshM	Boardman, 61, 44, 45, no 207
1544	Psy	0	1	2	7	1	0	45	3	0	0	0	1	7.4	0	0	0	0	0	0	0	0	AshM	Boardman, 61, 44, 45, no 204
1549	Psy	0	1	2	7	1	0	45	3	0	0	0	1	7.3	0	0	0	0	0	0	0	0	AshM	Boardman, 61, 44, 45, no 209
1541	Psy	0	1	2	7	1	0	45	3	0	0	0	1	15	0	0	0	0	0	0	0	0	AshM	Boardman, 61, 44, 45, no 201
1546	Psy	0	1	2	7	1	0	45	3	0	0	0	1	18	0	0	0	0	0	0	0	0	AshM	Boardman, 61, 44, 45, no 206

Appendix V.2 Bronze objects. Sorted by object type / period / site. Total 1857

No	Site	Co	G	L	S	Pe	Da	Ot	C	TM	TK	TP	Cn	Le	Wi	He	Rd	We	Cu	Sn	As	Pb	Mu	Ref
1555	Psy	0	1	2	7	1	0	45	3	0	0	0	1	0	0	0	0	0	0	0	0	0	484	Boardman, 61, 42, no 2
1550	Psy	0	1	2	7	1	0	45	3	0	0	0	1	0	0	0	0	0	0	0	0	0	485	Boardman, 61, 42, no 1
1551	Psy	0	1	2	7	1	0	45	3	0	0	0	1	0	0	0	0	0	0	0	0	0	486	Boardman, 61, 42, no 1
1540	Psy	0	1	2	7	1	0	45	3	0	0	0	3	5.1	6	0	0	0	0	0	0	0	AshM	Boardman, 61, 44, 45, no 200
1539	Psy	0	1	2	7	1	0	45	3	0	0	0	1	9.2	0	0	0	0	0	0	0	0	AshM	Boardman, 61, 44, no 199
1542	Psy	0	1	2	7	1	0	45	3	0	0	0	1	7.6	0	0	0	0	0	0	0	0	AshM	Boardman, 61, 44, 45, no 202
1801	Zak	XXVIII	1	1	1	1	4	45	3	0	0	0	1	47	0	0	0	0	0	0	0	0	HerM	Platon, 74, 143
1176	Iso	Tomb 2	2	2	6	2	5	45	3	0	0	0	3	0	0	0	0	0	0	0	0	0	0	Evans, 14, 58, no 2g
1175	Iso	Tomb 2	2	2	6	2	5	45	3	0	0	0	1	19	0	0	0	0	0	0	0	0	0	Evans, 14, 58, no 2g
1064	0	0	0	0	0	0	0	46	3	0	0	0	1	0	0	7.1	0	0	0	0	0	0	MeC	Sapouna-Sakellarakis, 95, 91, no 156
1063	0	0	0	0	0	0	0	46	3	0	0	0	1	0	0	6	0	0	0	0	0	0	MeC	Sapouna-Sakellarakis, 95, 91, no 155
1076	0	0	0	0	0	0	0	46	3	0	0	0	1	0	0	14	0	0	0	0	0	0	OrtC	Sapouna-Sakellarakis, 95, 96, no 168
1073	0	0	0	0	0	0	0	46	3	0	0	0	2	0	0	7	0	0	0	0	0	0	RoM	Sapouna-Sakellarakis, 95, 95, no 165
1077	0	0	0	0	0	0	0	46	3	0	0	0	2	0	0	17	0	0	0	0	0	0	MeM	Sapouna-Sakellarakis, 95, 96, no 169
1049	HaT	0	3	2	0	0	0	46	3	0	0	0	3	0	0	3.2	0	0	0	0	0	0	4528	Sapouna-Sakellarakis, 95, 75, no 128
1047	HaT	0	3	2	0	0	0	46	3	0	0	0	2	0	0	5	0	0	0	0	0	0	2312	Sapouna-Sakellarakis, 95, 74, no 126
1012	Iuk	0	2	2	8	0	0	46	3	0	0	0	1	0	0	6.3	0	0	0	0	0	0	4911	Sapouna-Sakellarakis, 95, 50, no 88
1011	Iuk	0	2	2	8	0	0	46	3	0	0	0	3	0	0	2.5	0	0	0	0	0	0	4450	Sapouna-Sakellarakis, 95, 50, no 87
1053	Pha	0	3	2	2	0	0	46	3	0	0	0	2	0	0	14	0	0	0	0	0	0	LeiM	Sapouna-Sakellarakis, 95, 77, no 133
963	Psy	0	1	2	7	0	0	46	3	0	0	0	3	0	0	5	0	0	0	0	0	0	1817	Sapouna-Sakellarakis, 95, 27, no 31
977	Psy	0	1	2	7	0	0	46	3	0	0	0	3	0	0	5.5	0	0	98.5	0	0	0.6	AshM	Sapouna-Sakellarakis, 95, 32, no 46, 153
990	Sym	0	1	2	8	0	0	46	3	0	0	0	2	0	0	5.4	0	0	0	0	0	0	0	Sapouna-Sakellarakis, 95, 37, no 59
988	Sym	Area1 B	1	2	8	0	0	46	3	0	0	0	1	0	0	10	0	0	0	0	0	0	0	Sapouna-Sakellarakis, 95, 37, no 57
987	Sym	Area1 B	1	2	8	0	0	46	3	0	0	0	1	0	0	9.1	0	0	0	0	0	0	0	Sapouna-Sakellarakis, 95, 36, no 56
989	Sym	0	1	2	8	0	0	46	3	0	0	0	1	0	0	7	0	0	0	0	0	0	0	Sapouna-Sakellarakis, 95, 37, no 58
1072	0	0	0	0	0	1	0	46	3	0	0	0	1	0	0	18	0	0	96	3	0	1	SaOM	Sapouna-Sakellarakis, 95, 94, no 164, 153
1062	0	0	0	0	0	1	0	46	3	0	0	0	1	0	0	7	0	0	0	0	0	0	GiaC	Sapouna-Sakellarakis, 95, 90, no 154
1070	0	0	0	0	0	1	0	46	3	0	0	0	1	0	0	22	0	0	98	0.1	0	0	LonM	Sapouna-Sakellarakis, 95, 93, no 162, 153
1069	0	0	0	0	0	1	0	46	3	0	0	0	1	0	0	9	0	0	0	0	0	0	FwM	Sapouna-Sakellarakis, 95, 93, no 161
1065	0	0	0	0	0	1	0	46	3	0	0	0	1	0	0	10	0	0	94.8	0	0	4.9	MeC	Sapouna-Sakellarakis, 95, 91, no 157, 153
1071	0	0	0	0	0	1	0	46	3	0	0	0	1	0	0	12	0	0	0	0	0	0	MaM	Sapouna-Sakellarakis, 95, 94, no 163
1066	0	0	0	0	0	1	0	46	3	0	0	0	1	0	0	7	0	0	0	0	0	0	AthM	Sapouna-Sakellarakis, 95, 92, no 158
1075	0	0	0	0	0	1	0	46	3	0	0	0	2	0	0	5.3	0	0	0	0	0	0	OrtC	Sapouna-Sakellarakis, 95, 96, no 167
1068	0	0	0	0	0	1	0	46	3	0	0	0	1	0	0	7.2	0	0	95.4	1.3	1.2	0.1	BerM	Sapouna-Sakellarakis, 95, 93, no 160, 153
1067	0	0	0	0	0	1	0	46	3	0	0	0	2	0	0	7.5	0	0	0	0	0	0	BerM	Sapouna-Sakellarakis, 95, 92, no 159
1003	Arh	Tourk./ 3	2	2	4	1	3	46	3	0	0	0	3	0	0	1.5	0	0	0	0	0	0	5189	Sapouna-Sakellarakis, 95, 46, no 75
1005	Arh	0	2	2	3	1	0	46	3	0	0	0	1	0	0	11	0	0	0	0	0	0	OrtC	Sapouna-Sakellarakis, 95, 46, no 77
1002	Arh	Hag.Triada	2	2	3	1	0	46	3	0	0	0	1	0	0	7	0	0	98.6	0	0	0.6	2508	Sapouna-Sakellarakis, 95, 45, no 74, 157
1004	Arh	Troullos	2	2	3	1	0	46	3	0	0	0	3	0	0	3	0	0	0	0	0	0	5188	Sapouna-Sakellarakis, 95, 46, no 76
1061	Cre	0	0	0	0	1	0	46	3	0	0	0	3	0	0	4	0	0	0	0	0	0	MeM	Sapouna-Sakellarakis, 95, 83, no 143

Appendix V.2 Bronze objects. Sorted by object type / period / site. Total 1857

No	Site	Co	G	L	S	Pe	Da	Ot	C	TM	TK	TP	Cn	Le	Wi	He	Rd	We	Cu	Sn	As	Pb	Mu	Ref
1059	Cre	0	0	0	0	1	0	46	3	0	0	0	1	0	0	9.7	0	0	0	0	0	0	WieM	Sapouna-Sakellarakis, 95, 82, no 141
944	Elo	0	1	1	0	1	0	46	3	0	0	0	1	0	0	10	0	0	0	0	0	0	GiaC	Sapouna-Sakellarakis, 95, 16, no 12
943	Gou	G 27	1	1	3	1	4	46	3	0	0	0	1	0	0	11	0	0	95	4	0.1	0.2	612	Sapouna-Sakellarakis, 95, 14, no 10, 155, 159
1034	Grv	0	4	2	8	1	0	46	3	0	0	0	1	0	0	26	0	0	0	0	0	0	2314	Sapouna-Sakellarakis, 95, 66, no 113
1055	HaT	Piaz. d. Sac.	3	2	4	1	0	46	3	0	0	0	2	0	0	10	0	0	0	0	0	0	PigM	Sapouna-Sakellarakis, 95, 78, no 135
1054	HaT	Piaz. d. Sac.	3	2	4	1	0	46	3	0	0	0	1	0	0	8.5	0	0	0	0	0	0	PigM	Sapouna-Sakellarakis, 95, 78, no 134
1037	HaT	Piaz. d. Sac.	3	2	4	1	0	46	3	0	0	0	1	0	0	6	0	0	0	0	0	0	750	Sapouna-Sakellarakis, 95, 69, no 116, 68
1050	HaT	0	3	2	0	1	0	46	3	0	0	0	3	0	0	6.8	0	0	87.6	11.6	0	0.3	4781	Sapouna-Sakellarakis, 95, 75, no 129, 157
1046	HaT	Tomb area	3	2	6	1	0	46	3	0	0	0	1	0	0	13	0	0	52.7	0.2	0	46.	2054	Sapouna-Sakellarakis, 95, 74, no 125, 155
1038	HaT	Piaz. d. Sac.	3	2	4	1	0	46	3	0	0	0	1	0	0	8.5	0	0	0	0	0	0	752	Sapouna-Sakellarakis, 95, 69, no 117
1036	HaT	Piaz. d. Sac.	3	2	4	1	0	46	3	0	0	0	3	0	0	8.8	0	0	0	0	0	0	749	Sapouna-Sakellarakis, 95, 69, no 115
1052	HaT	0	3	2	0	1	0	46	3	0	0	0	3	0	0	3.1	0	0	88.2	10.9	0	0.3	4783	Sapouna-Sakellarakis, 95, 76, no 131, 157
1051	HaT	0	3	2	0	1	0	46	3	0	0	0	3	0	0	3	0	0	86.9	0.1	0	12.	4782	Sapouna-Sakellarakis, 95, 76, no 130, 155
1008	Iuk	0	2	2	8	1	0	46	3	0	0	0	3	0	0	2.1	0	0	0	0	0	0	4231	Sapouna-Sakellarakis, 95, 49, no 83
1009	Iuk	0	2	2	8	1	0	46	3	0	0	0	3	0	0	2.5	0	0	72.5	0	0	25.	4273	Sapouna-Sakellarakis, 95, 50, no 85, 157
1013	Iuk	0	2	2	8	1	0	46	3	0	0	0	1	0	0	8.9	0	0	98.6	0	1.2	0.2	4912	Sapouna-Sakellarakis, 95, 51, no 89, 160
1010	Iuk	0	2	2	8	1	0	46	3	0	0	0	1	0	0	3.9	0	0	0	0	0	0	4432	Sapouna-Sakellarakis, 95, 50, no 86
1007	Iuk	0	2	2	8	1	0	46	3	0	0	0	3	0	0	4.1	0	0	0	0	0	0	3836	Sapouna-Sakellarakis, 95, 48, no 80
1019	Kat	0	2	1	0	1	0	46	3	0	0	0	1	0	0	12	0	0	0	0	0	0	OrtC	Sapouna-Sakellarakis, 95, 57, no 98
1018	Kat	0	2	1	6	1	0	46	3	0	0	0	3	0	0	11	0	0	95.2	1	0.8	1.9	1829	Sapouna-Sakellarakis, 95, 55, no 97, 157
1020	Kat	0	2	1	3	1	0	46	3	0	0	0	1	0	0	9.3	0	0	0	0	0	0	2278	Sapouna-Sakellarakis, 95, 57, no 99
1015	Kno	0	2	2	0	1	0	46	3	0	0	0	1	0	0	6.3	0	0	0	0	0	0	639	Sapouna-Sakellarakis, 95, 53, no 91
1016	Kno	0	2	2	0	1	0	46	3	0	0	0	1	0	0	11	0	0	0	0	0	0	WieM	Sapouna-Sakellarakis, 95, 54, no 94
999	Kop	0	1	2	8	1	0	46	3	0	0	0	1	0	0	7.3	0	0	0	0	0	0	MeC	Sapouna-Sakellarakis, 95, 43, no 70
997	Kop	0	1	2	8	1	0	46	3	0	0	0	2	0	0	5.2	0	0	0	0	0	0	MeC	Sapouna-Sakellarakis, 95, 42, no 68
998	Kop	0	1	2	8	1	0	46	3	0	0	0	2	0	0	3.4	0	0	0	0	0	0	MeC	Sapouna-Sakellarakis, 95, 43, no 69
992	Mal	0	2	1	0	1	0	46	3	0	0	0	1	0	0	11	0	0	0	0	0	0	GiaC	Sapouna-Sakellarakis, 95, 38, no 61
978	Psy	0	1	2	7	1	0	46	3	0	0	0	2	0	0	5.8	0	0	0	0	0	0	AshM	Sapouna-Sakellarakis, 95, 33, no 47
972	Psy	0	1	2	7	1	0	46	3	0	0	0	1	0	0	7	0	0	100	0	0	0	AshM	Sapouna-Sakellarakis, 95, 30, no 41, 153
983	Psy	0	1	2	7	1	0	46	3	0	0	0	3	0	0	3.2	0	0	0	0	0	0	AshM	Sapouna-Sakellarakis, 95, 34, no 52
982	Psy	0	1	2	7	1	0	46	3	0	0	0	3	0	0	2.3	0	0	0	0	0	0	AshM	Sapouna-Sakellarakis, 95, 34, no 51
949	Psy	0	1	2	7	1	0	46	3	0	0	0	1	0	0	5.5	0	0	0	0	0	0	426	Sapouna-Sakellarakis, 95, 21, no 17
984	Psy	0	1	2	7	1	0	46	3	0	0	0	1	0	0	18	0	0	99	0	0	0	Louv	Sapouna-Sakellarakis, 95, 34, no 53, 153
962	Psy	0	1	2	7	1	0	46	3	0	0	0	1	0	0	8.3	0	0	0	0	0	0	1814	Sapouna-Sakellarakis, 95, 26, no 30
957	Psy	0	1	2	7	1	0	46	3	0	0	0	2	0	0	4.5	0	0	76.3	0.4	0.2	23	435	Sapouna-Sakellarakis, 95, 24, no 25, 155
961	Psy	0	1	2	7	1	0	46	3	0	0	0	3	0	0	3.5	0	0	0	0	0	0	514	Sapouna-Sakellarakis, 95, 26, no 29
968	Psy	0	1	2	7	1	0	46	3	0	0	0	3	0	0	4.9	0	0	98	1	0.1	0.1	4780	Sapouna-Sakellarakis, 95, 29, no 37, 155
966	Psy	0	1	2	7	1	0	46	3	0	0	0	3	0	0	5	0	0	60.6	0	0.2	38.	4530	Sapouna-Sakellarakis, 95, 28, no 35, 155
970	Psy	0	1	2	7	1	0	46	3	0	0	0	1	0	0	3.2	0	0	0	0	0	0	GiaC	Sapouna-Sakellarakis, 95, 30, no 39
995	Sko	0	2	2	7	1	3	46	3	0	0	0	2	0	0	11	0	0	89.4	1.6	0.6	7.9	2575	Sapouna-Sakellarakis, 95, 40, no 64, 157

Appendix V.2 Bronze objects. Sorted by object type / period / site. Total 1857

No	Site	Co	G	L	S	Pe	Da	Ot	C	TM	TK	TP	Cn	Le	Wi	He	Rd	We	Cu	Sn	As	Pb	Mu	Ref
994	Sko	0	2	2	7	1	3	46	3	0	0	0	2	0	0	7.7	0	0	96.3	0.5	1.1	0.2	2574	Sapouna-Sakellarakis, 95, 39, no 63, 158
993	Sko	0	2	2	7	1	3	46	3	0	0	0	2	0	0	9	0	0	98.1	0.2	1	0.1	2573	Sapouna-Sakellarakis, 95, 39, no 62, 158
1025	Tyl	0	4	2	0	1	0	46	3	0	0	0	1	0	0	6.5	0	0	0	0	0	0	MeC	Sapouna-Sakellarakis, 95, 62, no 104
1024	Tyl	0	4	2	0	1	0	46	3	0	0	0	1	0	0	6	0	0	0	0	0	0	MeC	Sapouna-Sakellarakis, 95, 61, no 103
1029	Tyl	Trap Cave	4	2	7	1	0	46	3	0	0	0	1	0	0	8.8	0	0	0	0	0	0	OrtC	Sapouna-Sakellarakis, 95, 64, no 108
1032	Tyl	Trap Cave	4	2	7	1	0	46	3	0	0	0	1	0	0	7	0	0	0	0	0	0	BarC	Sapouna-Sakellarakis, 95, 65, no 111
1023	Tyl	0	4	2	5	1	0	46	3	0	0	0	1	0	0	11	0	0	97.7	1.5	0.2	0	1832	Sapouna-Sakellarakis, 95, 61, no 102, 157
1022	Tyl	0	4	2	5	1	0	46	3	0	0	0	1	0	0	17	0	0	99.1	0.1	0.5	0	1831	Sapouna-Sakellarakis, 95, 60, no 101, 157
1021	Tyl	House A	4	2	4	1	0	46	3	0	0	0	1	0	0	20	0	0	97	1.7	0.4	0.1	1762	Sapouna-Sakellarakis, 95, 59, no 100, 156
1026	Tyl	0	4	2	0	1	0	46	3	0	0	0	1	0	0	3.7	0	0	0	0	0	0	MeC	Sapouna-Sakellarakis, 95, 62, no 105
1056	Vry	0	4	2	8	1	0	46	3	0	0	0	3	0	0	6	0	0	0	0	0	0	RetM	Sapouna-Sakellarakis, 95, 79, no 137
1079	Dre	0	1	2	3	2	5	46	3	0	0	0	2	0	0	7.5	0	0	88.3	8.5	1	0.1	2398	Sapouna-Sakellarakis, 95, 15, no 11, 158
1039	HaT	Piaz. d. Sac.3	2	2	4	2	5	46	3	0	0	0	2	0	0	3.6	0	0	70.9	0.5	0	27.	753	Sapouna-Sakellarakis, 95, 70, no 118, 157
1033	IdC	0	4	2	7	2	5	46	3	0	0	0	3	0	0	7	0	0	81.6	10.8	0.4	6.6	1641	Sapouna-Sakellarakis, 95, 66, no 112
1014	Kno	SW Wing	2	2	1	2	5	46	3	0	0	0	1	0	0	12	0	0	76.9	12.2	1.2	7.1	704	Sapouna-Sakellarakis, 95, 52, no 90, 158
991	Mal	Azymo	2	1	0	2	5	46	3	0	0	0	2	0	0	7.1	0	0	73.4	2.1	0.7	23	2958	Sapouna-Sakellarakis, 95, 37, no 60, 157
940	Pak	Block X	1	1	3	2	5	46	3	0	0	0	3	0	0	3.5	0	0	94.6	0.3	0.7	0.1	1418	Sapouna-Sakellarakis, 95, 10, no 2. 158, 160
956	Psy	0	1	2	7	2	5	46	3	0	0	0	1	0	0	6	0	0	0	0	0	0	434	Sapouna-Sakellarakis, 95, 24, no 24
965	Psy	0	1	2	7	2	5	46	3	0	0	0	1	0	0	7.5	0	0	98.8	0	0.2	0	4529	Sapouna-Sakellarakis, 95, 28, no 34, 155
985	Psy	0	1	2	7	2	5	46	3	0	0	0	3	0	0	8.3	0	0	0	0	0	0	MeM	Sapouna-Sakellarakis, 95, 35, no 54
948	Psy	0	1	2	7	2	5	46	3	0	0	0	1	0	0	7.5	0	0	0	0	0	0	425	Sapouna-Sakellarakis, 95, 20, no 16
981	Psy	0	1	2	7	2	5	46	3	0	0	0	3	0	0	5	0	0	99.5	0	0.4	0.1	AshM	Sapouna-Sakellarakis, 95, 34, no 50, 153
945	Psy	0	1	2	7	2	5	46	3	0	0	0	1	0	0	8.5	0	0	97.2	0	1	0.2	204	Sapouna-Sakellarakis, 95, 18, no 13, 155
947	Psy	0	1	2	7	2	5	46	3	0	0	0	1	0	0	7	0	0	0	0	0	0	423	Sapouna-Sakellarakis, 95, 20, no 15
955	Psy	0	1	2	7	2	5	46	3	0	0	0	1	0	0	7	0	0	0	0	0	0	432	Sapouna-Sakellarakis, 95, 23, no 23
1006	Arh	0	2	2	3	3	0	46	3	0	0	0	2	0	0	4.2	0	0	0	0	0	0	4351	Sapouna-Sakellarakis, 95, 47, no 78
1058	Cre	0	0	0	0	3	0	46	3	0	0	0	1	0	0	24	0	0	98.1	0	0.3	0	BerM	Sapouna-Sakellarakis, 95, 81, no 140, 153
973	Psy	0	1	2	7	3	0	46	3	0	0	0	1	0	0	13	0	0	99	0.4	0.1	0.3	AshM	Sapouna-Sakellarakis, 95, 31, no 42, 153
946	Psy	0	1	2	7	3	0	46	3	0	0	0	1	0	0	12	0	0	0	0	0	0	205	Sapouna-Sakellarakis, 95, 19, no 14
954	Psy	0	1	2	7	3	0	46	3	0	0	0	2	0	0	9	0	0	0	0	0	0	431	Sapouna-Sakellarakis, 95, 23, no 22
953	Psy	0	1	2	7	3	0	46	3	0	0	0	1	0	0	12	0	0	99.4	0	0.3	0	430	Sapouna-Sakellarakis, 95, 23, no 21, 155
1028	Tyl	Trap Cave	4	2	7	1	0	47	3	0	0	0	1	0	0	4.2	0	0	0	0	0	0	OrtC	Sapouna-Sakellarakis, 95, 63, no 107
1027	Tyl	0	4	2	0	0	0	47	3	0	0	0	1	0	0	3.7	0	0	0	0	0	0	MeC	Sapouna-Sakellarakis, 95, 63, no 106
1078	0	0	0	0	0	1	0	47	3	0	0	0	1	0	0	6.9	0	0	0	0	0	0	CaM	Sapouna-Sakellarakis, 95, 96, no 170
1074	0	0	0	0	0	1	0	47	3	0	0	0	2	0	0	6.7	0	0	0	0	0	0	OrtC	Sapouna-Sakellarakis, 95, 95, no 166
1060	Cre	0	0	0	0	1	0	47	3	0	0	0	1	0	0	4.5	0	0	0	0	0	0	BosM	Sapouna-Sakellarakis, 95, 82, no 142
996	EpP	0	2	2	0	1	0	47	3	0	0	0	1	0	0	3.7	0	0	0	0	0	0	MeM	Sapouna-Sakellarakis, 95, 41, no 66
1042	HaT	Piaz. d. Sac.3	2	2	4	1	0	47	3	0	0	0	1	0	0	4	0	0	99.3	0.2	0.2	0.1	759	Sapouna-Sakellarakis, 95, 71, no 121, 157
1048	HaT	0	3	2	0	1	0	47	3	0	0	0	3	0	0	4	0	0	85.3	0.2	0.2	13.	4527	Sapouna-Sakellarakis, 95, 75, no 127, 157
1040	HaT	Piaz. d. Sac.3	2	2	4	1	0	47	3	0	0	0	1	0	0	5.6	0	0	0	0	0	0	756	Sapouna-Sakellarakis, 95, 70, no 119

Appendix V.2 Bronze objects. Sorted by object type / period / site. Total 1857

No	Site	Co	G	L	S	Pe	Da	Ot	C	TM	TK	TP	Cn	Le	Wi	He	Rd	We	Cu	Sn	As	Pb	Mu	Ref
1044	HaT	Piaz. d. Sac.3	2	2	4	1	0	47	3	0	0	0	1	0	0	7.5	0	0	0	0	0	0	761	Sapouna-Sakellarakis, 95, 73, no 123, 68
1045	HaT	Piaz. d. Sac.3	2	2	4	1	0	47	3	0	0	0	3	0	0	2.5	0	0	92.7	0.1	0.5	5.3	1029	Sapouna-Sakellarakis, 95, 73, no 124, 157
1043	HaT	Piaz. d. Sac.3	2	2	4	1	0	47	3	0	0	0	2	0	0	14	0	0	99.3	0.2	0.1	0	760	Sapouna-Sakellarakis, 95, 72, no 122, 155
1000	Kop	0	2	2	8	1	0	47	3	0	0	0	1	0	0	4.8	0	0	0	0	0	0	MeC	Sapouna-Sakellarakis, 95, 43, no 71
1001	Kop	0	2	2	8	1	0	47	3	0	0	0	1	0	0	4.8	0	0	0	0	0	0	MeC	Sapouna-Sakellarakis, 95, 44, no 72
1035	Kop	0	2	2	8	1	0	47	3	0	0	0	1	0	0	4.6	0	0	0	0	0	0	MeC	Sapouna-Sakellarakis, 95, 67, no 114
942	Mak	0	1	1	4	1	4	47	3	0	0	0	1	0	0	6.2	0	0	0	0	0	0	HNM	Sapouna-Sakellarakis, 95, 14, no 9
939	Pak	Block X	1	1	3	1	0	47	3	0	0	0	1	0	0	4.4	0	0	96.9	0	0.9	0.1	1417	Sapouna-Sakellarakis, 95, 9, no 1, 158, 160
974	Psy	0	1	2	7	1	0	47	3	0	0	0	1	0	0	7.2	0	0	99	0.8	0.3	0.1	AshM	Sapouna-Sakellarakis, 95, 31, no 43, 153
958	Psy	0	1	2	7	1	0	47	3	0	0	0	1	0	0	5	0	0	84.1	0.9	0.1	12.	437	Sapouna-Sakellarakis, 95, 25, no 26, 155
960	Psy	0	1	2	7	1	0	47	3	0	0	0	1	0	0	4	0	0	0	0	0	0	439	Sapouna-Sakellarakis, 95, 25, no 28
959	Psy	0	1	2	7	1	4	47	3	0	0	0	1	0	0	4	0	0	0	0	0	0	438	Sapouna-Sakellarakis, 95, 25, no 27
969	Psy	0	1	2	7	1	0	47	3	0	0	0	1	0	0	4.1	0	0	0	0	0	0	GiaC	Sapouna-Sakellarakis, 95, 29, no 38
986	Psy	0	1	2	7	1	0	47	3	0	0	0	1	0	0	5.5	0	0	0	0	0	0	GiaC	Sapouna-Sakellarakis, 95, 36, no 55
952	Psy	0	1	2	7	1	0	47	3	0	0	0	1	0	0	6.5	0	0	97.6	0	0.6	0.2	429	Sapouna-Sakellarakis, 95, 22, no 20, 155
971	Psy	0	1	2	7	1	0	47	3	0	0	0	1	0	0	7.8	0	0	93	3	1.2	0.6	AshM	Sapouna-Sakellarakis, 95, 30, no 40, 153
967	Psy	0	1	2	7	1	0	47	3	0	0	0	3	0	0	2.6	0	0	72.8	0.4	0.4	25.	4779	Sapouna-Sakellarakis, 95, 28, no 36, 155
975	Psy	0	1	2	7	1	0	47	3	0	0	0	1	0	0	5.3	0	0	74.5	0	0	24.	AshM	Sapouna-Sakellarakis, 95, 32, no 44, 153
1030	Tyl	Trap Cave	4	2	7	1	0	47	3	0	0	0	1	0	0	5.5	0	0	0	0	0	0	OrtC	Sapouna-Sakellarakis, 95, 64, no 109
1031	Tyl	Trap Cave	4	2	7	1	0	47	3	0	0	0	1	0	0	4.8	0	0	0	0	0	0	OrtC	Sapouna-Sakellarakis, 95, 64, no 110
1041	HaT	Piaz. d. Sac.3	2	2	4	2	5	47	3	0	0	0	1	0	0	4.5	0	0	97.2	1.6	0.3	0.5	758	Sapouna-Sakellarakis, 95, 71, no 120, 155
1017	Kno	UM Ro 27	2	2	2	2	5	47	3	0	0	0	1	0	0	4.6	0	0	0	0	0	0	3133	Sapouna-Sakellarakis, 95, 54, no 95
941	Pak	Block X	1	1	3	2	5	47	3	0	0	0	1	0	0	2	0	0	0	0	0	0	1419	Sapouna-Sakellarakis, 95, 11, no 3
964	Psy	0	1	2	7	2	5	47	3	0	0	0	1	0	0	3.8	0	0	0	0	0	0	1833	Sapouna-Sakellarakis, 95, 27, no 32
950	Psy	0	1	2	7	3	5	47	3	0	0	0	1	0	0	8	0	0	0	0	0	0	427	Sapouna-Sakellarakis, 95, 21, no 18
951	Psy	0	1	2	7	3	0	47	3	0	0	0	1	0	0	6	0	0	0	0	0	0	428	Sapouna-Sakellarakis, 95, 22, no 19
979	Psy	0	1	2	7	3	0	47	3	0	0	0	1	0	0	3.3	0	0	99	0	0.9	0.2	AshM	Sapouna-Sakellarakis, 95, 33, no 48, 153
976	Psy	0	1	2	7	3	0	47	3	0	0	0	1	0	0	4.7	0	0	70.5	1.2	1	27.	AshM	Sapouna-Sakellarakis, 95, 32, no 45, 153
980	Psy	0	1	2	7	1	0	49	3	0	0	0	1	4.8	0	2.8	0	0	99.5	0	0.4	0.1	AshM	Sapouna-Sakellarakis, 95, 33, no 49, 153
1057	Ret	0	4	1	0	1	0	49	3	0	0	0	2	0	0	11	0	0	96.4	1.5	0.5	0.7	LonM	Sapouna-Sakellarakis, 95, 79, no 138,153
1169	Iso	Royal Tomb2	2	2	6	1	0	51	2	0	0	0	1	0	0	0	22	0	0	0	0	0	0	Evans, 06, 155, no 36, 170
1721	Ath	Tholos A	2	2	6	2	6	51	2	0	0	0	1	0	0	0	0	0	0	0	0	0	HerM	Sakellarakis / Sakellarakis, 97, 602
1465	Cha	0	4	1	6	2	6	51	2	0	0	0	1	0	0	0	13	0	0	0	0	0	0	Kanta, 80, 224
1466	Cha	0	4	1	6	2	6	51	2	0	0	0	1	0	0	0	0	0	0	0	0	0	0	Kanta, 80, 225
1468	Cha	0	4	1	6	2	6	51	2	0	0	0	1	0	0	0	0	0	0	0	0	0	0	Kanta, 80, 226
1467	Cha	0	4	1	6	2	6	51	2	0	0	0	1	0	0	0	0	0	0	0	0	0	0	Kanta, 80, 226
1180	Iso	Tomb 3	2	2	6	2	6	51	2	0	0	0	1	0	0	0	0	0	0	0	0	0	0	Evans, 14, 15, no 3f
1181	Iso	Tomb 6	2	2	6	2	6	51	2	0	0	0	3	0	0	0	8	0	0	0	0	0	0	Evans, 14, 33, no 6n
1203	Mas	Tomb III	2	2	6	2	6	51	1	0	0	0	3	0	0	0	0	0	0	0	0	0	0	Forsdyke, 26-27, 252, no III.7
1227	Mas	Tomb XV	2	2	6	2	6	51	2	0	0	0	1	0	0	0	0	0	0	0	0	0	0	Forsdyke, 26-27, 275, no XV.1

Appendix V.2 Bronze objects. Sorted by object type / period / site. Total 1857

No	Site	Co	G	L	S	Pe	Da	Ot	C	TM	TK	TP	Cn	Le	Wi	He	Rd	We	Cu	Sn	As	Pb	Mu	Ref
1225	Mas	Tomb IXD	2	2	6	2	6	51	2	0	0	0	1	0	0	0	0	0	0	0	0	0	0	Forsdyke, 26-27, 268, no IXD.2
1237	Mas	Tomb XIX	2	2	6	2	6	51	2	0	0	0	1	0	0	0	0	0	0	0	0	0	0	Forsdyke, 26-27, 282, no XIX.1
1236	Mas	Tomb XVIII	2	2	6	2	6	51	2	0	0	0	1	0	0	0	0	0	0	0	0	0	0	Forsdyke, 26-27, 282, no XVIII.5
1189	Sel	Tomb 4/I	2	2	6	2	6	51	2	0	0	0	2	0	0	0	16	0	0	10.6	0	0	0	Catling / Catling, 74,230, no 13;Catling / J.,76, 22, no10
1190	Sel	Tomb 4/I	2	2	6	2	6	51	2	0	0	0	2	0	0	0	16	0	0	0	0	0	0	Catling / Catling, 74, 230, no 14
1192	Sel	Tomb 4/III	2	2	6	2	6	51	2	0	0	0	1	0	0	0	15	0	0	0	0	0	0	Catling / Catling, 74, 230, no 16
1191	Sel	Tomb 4/I	2	2	6	2	6	51	2	0	0	0	3	0	0	0	0	0	0	0	0	0	0	Catling / Catling, 74, 230, no 15
1193	Sel	Tomb 4/III	2	2	6	2	6	51	2	0	0	0	1	0	0	0	15	0	0	8.4	0	0	0	Catling / Catling, 74,230, no 17;Catling / J.,76, 22, no11
1198	Sel	Tomb 3	2	2	6	2	6	51	2	0	0	0	3	0	0	0	0	0	0	0	0	0	0	Catling / Catling, 74, 240, no 7
1162	ZaP	Tomb 95	2	2	6	2	6	51	2	0	0	0	1	14	0	0	0	0	0	0	0	0	0	Evans, 06, 84, no 95c
1111	ZaP	Tomb 7	2	2	6	2	6	51	2	0	0	0	1	13	0	0	0	0	0	0	0	0	0	Evans, 06, 25, no 7d
1143	ZaP	Tomb 67	2	2	6	2	6	51	2	0	0	0	1	0	0	0	13	0	0	0	0	0	0	Evans, 06, 73, no 67f
1142	ZaP	Tomb 66	2	2	6	2	6	51	2	0	0	0	1	0	0	0	0	0	0	0	0	0	0	Evans, 06, 72, no 66g
1155	ZaP	Tomb 76	2	2	6	2	6	51	2	0	0	0	2	0	0	0	13	0	0	0	0	0	0	Evans, 06, 78, no 76h
1128	ZaP	Tomb 36	2	2	6	2	6	51	2	0	0	0	1	18	0	0	0	0	0	0	0	0	0	Evans, 06, 55, no 36d
1163	ZaP	Tomb 95	2	2	6	2	6	51	2	0	0	0	1	0	0	0	0	0	0	0	0	0	0	Evans, 06, 85, no 95g
1109	ZaP	Tomb 6	2	2	6	2	6	51	2	0	0	0	2	0	0	0	17	0	0	0	0	0	0	Evans, 06, 25, no 6g
1166	ZaP	Tomb 99	2	2	6	2	6	51	2	0	0	0	1	0	0	0	13	0	0	0	0	0	AshM	Evans, 06, 90, no 99k; Baboula / Northover, 99, 150
1108	ZaP	Tomb 6	2	2	6	2	6	51	2	0	0	0	1	0	0	0	12	0	0	0	0	0	0	Evans, 06, 25, no 6f
1715	Arh	Tholos D	2	2	6	3	7	51	2	0	0	0	1	0	0	0	0	0	0	0	0	0	HerM	Sakellarakis / Sakellarakis, 97, 602
1722	Arh	Tholos D	2	2	6	3	7	51	2	0	0	0	1	0	0	0	0	0	0	0	0	0	HerM	Sakellarakis / Sakellarakis, 97, 602
1418	EkP	Tomb A	2	2	6	3	0	51	2	0	0	0	1	0	0	0	0	0	0	0	0	0	0	Kanta, 80, 61
1419	EkP	Potamo L	2	2	6	3	0	51	2	0	0	0	1	0	0	0	0	0	0	0	0	0	0	Kanta, 80, 65
1401	Gor	Tomb 4	2	2	6	3	8	51	2	0	0	0	1	0	0	0	0	0	0	0	0	0	0	Kanta, 80, 48
1400	Gor	Tomb 4	2	2	6	3	8	51	2	0	0	0	1	0	0	0	0	0	0	0	0	0	0	Kanta, 80, 48
1262	Gyp	Tomb X	2	2	6	3	8	51	2	0	0	0	1	0	0	0	14	0	0	0	0	0	0	Hood et al, 58-59, 250, no X.3
1404	KaE	0	2	2	6	3	0	51	2	0	0	0	1	0	0	0	0	0	0	0	0	0	0	Kanta, 80, 186
1763	Moc	Tomb 10	1	1	6	3	0	51	2	0	0	0	1	0	0	0	11	0	0	0	0	0	0	Soles / Davaras, 96, 212, no CA 97
1840	Moc	Tomb 10	1	1	6	3	8	51	2	0	0	0	1	0	0	0	0	0	0	0	0	0	0	Soles / Davaras, 96, 212
1432	Moi	0	3	2	6	3	7	51	2	0	0	0	1	0	0	0	0	0	0	0	0	0	0	Kanta, 80, 89
1442	Myr	Tomb H	1	1	6	3	7	51	2	0	0	0	1	0	0	0	0	0	0	0	0	0	0	Kanta, 80, 164, 172
1439	Paa	Pithos Bur	1	1	6	3	7	51	2	0	0	0	3	0	0	0	0	0	0	0	0	0	0	Kanta, 80, 143
1456	Phk	0	1	2	6	3	7	51	2	0	0	0	1	0	0	0	0	0	0	0	0	0	0	Kanta, 80, 164, 183
1270	Kep	0	2	2	2	1	6	52	2	0	0	0	1	2.1	0	0	0	0	0	0	0	0	0	Hutchinson, 56, 78, 79, no 10
1497	Mal	Quart D	2	1	2	1	0	52	2	0	0	0	1	0	0	0	0	0	0	0	0	0	0	Georgiou, 79, 29, no 11
1758	Moc	House C7	1	1	3	1	4	52	2	0	0	0	3	0	0	0	0	0	0	0	0	0	0	Soles / Davaras, 96, 201, no CA 128
1731	Moc	House A2	1	1	3	1	4	52	2	0	0	0	1	0	0	0	0	0	0	0	0	0	0	Soles / Davaras, 94, 419, no CA 27
1473	Cha	0	4	1	6	2	6	52	2	0	0	0	1	0	0	0	0	0	0	0	0	0	0	Kanta, 80, 225
1474	Cha	0	4	1	6	2	6	52	2	0	0	0	1	0	0	0	0	0	0	0	0	0	0	Kanta, 80, 225
1249	Gyp	Tomb IV	2	2	6	2	6	52	2	0	0	0	1	0	0	0	0	0	0	0	0	0	0	Hood et al, 58-59, 246, no IV.4

Appendix V.2 Bronze objects. Sorted by object type / period / site. Total 1857

No	Site	Co	G	L	S	Pe	Da	Ot	C	TM	TK	TP	Cn	Le	Wi	He	Rd	We	Cu	Sn	As	Pb	Mu	Ref
1247	Gyp	Tomb III	2	2	6	2	6	52	2	0	0	0	0	0	0	0	0	0	0	0	0	0	0	Hood et al, 58-59, 246, no III.1
1212	Mas	Tomb V	2	2	6	2	6	52	2	0	0	0	0	0	0	0	0	0	0	0	0	0	0	Forsdyke, 26-27, 257, no V.5
1220	Mas	Tomb VIIA	2	2	6	2	6	52	2	0	0	0	1	0	0	0	0	0	0	0	0	0	0	Forsdyke, 26-27, 262, no VIIA.5
1211	Mas	Tomb V	2	2	6	2	6	52	2	0	0	0	0	0	0	0	0	0	0	0	0	0	0	Forsdyke, 26-27, 257, no V.5
1219	Mas	Tomb VIIA	2	2	6	2	6	52	2	0	0	0	0	0	0	0	0	0	0	0	0	0	0	Forsdyke, 26-27, 262, no VIIA.4
1231	Mas	Tom XVIIB	2	2	6	2	6	52	2	0	0	0	1	0	0	0	0	0	0	0	0	0	0	Forsdyke, 26-27, 279, no XVIIB.1
1147	ZaP	Tomb 69	2	2	6	2	6	52	2	0	0	0	0	0	0	0	0	0	0	0	0	0	0	Evans, 06, 75
1146	ZaP	Tomb 69	2	2	6	2	6	52	2	0	0	0	0	0	0	0	0	0	0	0	0	0	0	Evans, 06, 75
1149	ZaP	Tomb 69	2	2	6	2	6	52	2	0	0	0	0	0	0	0	0	0	0	0	0	0	0	Evans, 06, 75
1167	ZaP	Tomb 100	2	2	6	2	6	52	2	0	0	0	1	0	0	0	0	0	0	0	0	0	0	Evans, 06, 92, no 100d
1148	ZaP	Tomb 69	2	2	6	2	6	52	2	0	0	0	0	0	0	0	0	0	0	0	0	0	0	Evans, 06, 75
1150	ZaP	Tomb 72	2	2	6	2	6	52	2	0	0	0	0	0	0	0	0	0	0	0	0	0	0	Evans, 06, 76, no 72a
1421	EkP	Khrios L	2	2	6	3	0	52	2	0	0	0	1	0	0	0	0	0	0	0	0	0	0	Kanta, 80, 66
1392	Gaz	0	2	1	6	3	8	52	2	0	0	0	1	0	0	0	0	0	0	0	0	0	0	Kanta, 80, 21
1393	Gaz	0	2	1	6	3	8	52	2	0	0	0	1	0	0	0	0	0	0	0	0	0	0	Kanta, 80, 21
1260	Gyp	Tomb VII	2	2	6	3	8	52	2	0	0	0	2	0	0	0	0	0	0	0	0	0	0	Hood et al, 58-59, 249, no VII.19
1259	Gyp	Tomb VII	2	2	6	3	8	52	2	0	0	0	3	0	0	0	0	0	0	0	0	0	0	Hood et al, 58-59, 249, no VII.18
1258	Gyp	Tomb VII	2	2	6	3	8	52	2	0	0	0	2	0	0	0	0	0	0	0	0	0	0	Hood et al, 58-59, 249, no VII.17
1766	Moc	Tomb 11	1	1	6	3	0	52	2	0	0	0	1	0	0	0	0	0	0	0	0	0	0	Soles / Davaras, 96, 216, no CA 91
1842	Moc	Tomb 11	1	1	6	3	8	52	2	0	0	0	1	0	0	0	0	0	0	0	0	0	0	Soles / Davaras, 96, 216
1767	Moc	Tomb 13	1	1	6	3	0	52	2	0	0	0	1	0	0	0	0	0	0	0	0	0	0	Soles / Davaras, 96, 218, no CA 100
1768	Moc	Tomb 13	1	1	6	3	0	52	2	0	0	0	1	0	0	0	0	0	0	0	0	0	0	Soles / Davaras, 96, 218, no CA 101
1457	Phk	0	1	2	6	3	7	52	2	0	0	0	1	0	0	0	0	0	0	0	0	0	0	Kanta, 80, 164, 183
1168	ZaP	Tomb 100	2	2	6	2	6	53	2	0	0	0	1	0	0	0	0	0	0	0	0	0	0	Evans, 06, 92, no 100e
1164	ZaP	Tomb 96	2	2	6	2	6	53	2	0	0	0	0	0	0	0	7.8	0	0	0	0	0	0	Evans, 06, 85, no 96a
1843	Moc	Tomb 11	1	2	6	3	8	53	2	0	0	0	0	0	0	0	0	0	0	0	0	0	0	Soles / Davaras, 96, 216
1765	Moc	Tomb 11	1	2	6	3	0	53	2	0	0	0	1	0	0	0	0	0	0	0	0	0	0	Soles / Davaras, 96, 216, no CA 89
1246	Gyp	Tomb II	2	2	6	2	6	54	2	0	0	0	1	0	0	0	1.4	0	0	0	0	0	0	Hood et al, 58-59, 246, no II.7
1218	Mas	Tomb VIIA	2	2	6	2	6	54	2	0	0	0	1	0	0	0	0	0	0	0	0	0	0	Forsdyke, 26-27, 262, no VIIA.6
1216	Mas	Tomb VIIA	2	2	6	2	6	54	2	0	0	0	1	0	0	0	0	0	0	0	0	0	0	Forsdyke, 26-27, 262, no VIIA.3
1217	Mas	Tomb VIIA	2	2	6	2	6	54	2	0	0	0	1	0	0	0	0	0	0	0	0	0	0	Forsdyke, 26-27, 262, no VIIA.3
1214	Mas	Tomb VIIA	2	2	6	2	6	54	2	0	0	0	1	0	0	0	0	0	0	0	0	0	0	Forsdyke, 26-27, 262, no VIIA.3
1213	Mas	Tomb V	2	2	6	2	6	54	2	0	0	0	1	0	0	0	0	0	0	0	0	0	0	Forsdyke, 26-27, 257, no V.6
1215	Mas	Tomb VIIA	2	2	6	2	6	54	2	0	0	0	1	0	0	0	0	0	0	0	0	0	0	Forsdyke, 26-27, 262, no VIIA.3
1204	Mas	Tomb IV	2	2	6	2	6	54	2	0	0	0	2	0	0	0	0	0	0	0	0	0	0	Forsdyke, 26-27, 255, no IV.2
1207	Mas	Tomb IV	2	2	6	2	6	54	2	0	0	0	2	0	0	0	0	0	0	0	0	0	0	Forsdyke, 26-27, 255, no IV.2
1206	Mas	Tomb IV	2	2	6	2	6	54	2	0	0	0	2	0	0	0	0	0	0	0	0	0	0	Forsdyke, 26-27, 255, no IV.2
1205	Mas	Tomb IV	2	2	6	2	6	54	2	0	0	0	2	0	0	0	0	0	0	0	0	0	0	Forsdyke, 26-27, 255, no IV.2
1251	Gyp	Tomb VI	2	2	6	3	8	54	2	0	0	0	1	17	0	0	1.1	0	0	0	0	0	0	Hood et al, 58-59, 246, no VI.3
1440	Paa	Pithos Bur	1	1	6	3	7	54	2	0	0	0	1	0	0	0	0	0	0	0	0	0	0	Kanta, 80, 143

Appendix V.2 Bronze objects. Sorted by object type / period / site. Total 1857

No	Site	Co	G	L	S	Pe	Da	Ot	C	TM	TK	TP	Cn	Le	Wi	He	Rd	We	Cu	Sn	As	Pb	Mu	Ref
1487	Kno	UM O	2	2	2	2	5	55	2	0	0	0	1	4.3	0	0	0	0	92	8	0	0	0	Popham, 84, 76, no O 2; Catling / Catling, 84, 217
1486	Kno	UM L	2	2	2	2	5	55	2	0	0	0	1	5.5	0	0	0	0	92.5	7.5	0	0	0	Popham, 84, 50, no L 120; Catling / Catling, 84, 217
1224	Mas	Tomb IXD	2	2	6	2	6	55	2	0	0	0	1	6.3	0	0	0	0	0	0	0	0	0	Forsdyke, 26-27, 268, no IXD.1
1285	NeH	Tomb V	2	2	6	2	6	55	2	0	0	0	1	5.5	0	0	0	0	0	0	0	0	0	Hood / d.Jong, 52, 277, no 10
1271	Kep	0	2	2	6	1	6	56	2	0	0	0	2	2.2	0	0	0	0	0	0	0	0	0	Hutchinson, 56, 78, 79, no 11
1723	Moc	House C2	1	1	3	1	4	56	2	0	0	0	1	0	0	0	0	0	0	0	0	0	0	Soles / Davaras, 94, 398, no CA 10
1233	Mas	Tom XVIIB	2	2	6	2	6	56	2	0	0	0	2	0	0	0	0	0	0	0	0	0	0	Forsdyke, 26-27, 279, no XVIIP.1
1230	Mas	Tom XVIIA	2	2	6	2	6	56	2	0	0	0	1	0	0	0	0	0	0	0	0	0	0	Forsdyke, 26-27, 278, no XVIIA.2
1229	Mas	Tom XVIIA	2	2	6	2	6	56	2	0	0	0	1	0	0	0	0	0	0	0	0	0	0	Forsdyke, 26-27, 278, no XVIIA.2
1261	Gyp	Tomb IX	2	2	6	3	8	56	2	0	0	0	1	0	0	0	0	0	0	0	0	0	0	Hood et al, 58-59, 250, no IX.3
1705	Gou	C 30	1	1	3	1	4	60	3	0	0	0	1	1.6	1.8	0	0	0	0	0	0	0	836	Boyd Hawes, 08, 48, no 16
1704	Gou	A 43	1	1	3	1	4	60	3	0	0	0	1	2.1	1.8	0	0	0	0	0	0	0	837	Boyd Hawes, 08, 48, no 15
1269	Kep	0	2	2	6	1	6	60	0	0	0	0	3	0	0	0	0	0	0	0	0	0	0	Hutchinson, 56, 78, 79, no 9
1276	Kep	0	2	2	6	1	6	60	0	0	0	0	3	4.5	0	0	0	0	0	0	0	0	0	Hutchinson, 56, 78, 79, no 16
1275	Kep	0	2	2	6	1	6	60	0	0	0	0	3	0	0	0	0	0	0	0	0	0	0	Hutchinson, 56, 78, 79, no 15
1278	Hal	Tomb 2	2	2	6	2	5	60	0	0	0	0	3	0	0	0	0	0	0	0	0	0	0	Hood, 56, 99, no 21
1282	NeH	Tomb III	2	2	6	2	6	60	0	0	0	0	3	0	0	0	0	0	0	0	0	0	0	Hood / d.Jong, 52, 271, no 16
1286	NeH	Tomb V	2	2	6	2	6	60	0	0	0	0	3	14	0	0	0	0	0	0	0	0	0	Hood / d.Jong, 52, 277, no 11

www.ingramcontent.com/pod-product-compliance
Lightning Source LLC
Chambersburg PA
CBHW061003030426

42334CB00033B/3342